IMAGES
of Wales

SOUTH WALES COLLIERIES

VOLUME FOUR

Children in the Mines

Words and music by Hawys Glyn James

1. My name is John, I'm seven years old
   And I'm a doorboy in the pit,
   My job's to open and close doors,
   So in the darkness there I sit,
   At six o'clock I leave my home,
   I'd really like to stay in bed,
   The day is long, I work till seven
   And big rats steal my cheese and bread.

Chorus
The Four-Foot Level there I work,
I don't shirk,
That's my work.
The Four-Foot Level there I work,
I am the air door keeper.

2. My brother, Tom, is ten years old,
   He drags the coal drams from the face,
   He has to crawl on knees and hands,
   The roof is low, there's little space.
   With leather strap around his waist
   And through his legs run two strong chains,
   He pulls his dram just like a horse
   And Tom earns 10p for his pains.

3. My Mam and sister work the winch
   That winds the coal drams up the slope,
   My brother, William, works with Dad
   Whose chest is bad so he can't cope,
   Last week a doorboy fell asleep,
   On to the rail he slowly rolled,
   A heavy journey crushed his bones
   And he was only six years old.

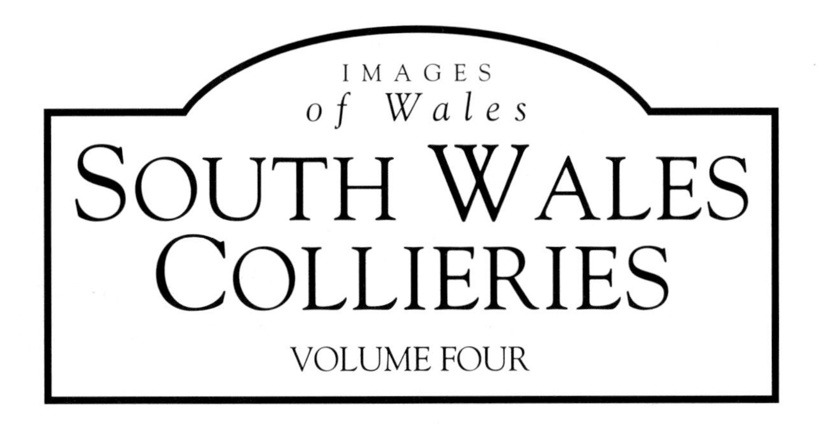

Valley and Vale

From the Central Valleys of Merthyr Tydfil, Glamorganshire to the Eastern Valleys of Rhymney, Sirhowy, Ebbw and Afon Lwyd, Monmouthshire in the far East of the South Wales Coalfield

David Owen

Acknowledgements

Thank you all for the wonderful stories, songs, poems, drawings and photographs of the South Wales Collieries, which have been given to me by the people from the mining villages of South Wales.

These have come from the early days of the coal industry through to the new millennium. I dedicate my book to the people of South Wales the Land of Song and in memory of all the miners who worked at the collieries.

I sincerely thank everyone for their kindness and help.

David Owen
Author and Archivist

Cydnabyddiaeth

Diolch am yr holl storïau, caneuon, cerddi, darluniau a ffotograffau aruthrol o Faes Glo De Cymru, sydd wedi eu cynnig i mi gan bobl pentrefi glofaol De Cymru.

Mae'r cyfraniadau yma yn dod o ddyddiau cynnar y diwydiant glo trwyddo i'r milflwydd newydd. Rwy'n cyflwyno'r llyfr yma i bobl Gwlad y Gân De Cymru er cof am y glowyr a wethiodd yn y pyllau glo.

Rwy'n diolch yn ddidwyll i bawb am eu caredigrwydd a cymorth.

David Owen
Awdur ac Archifydd

First published 2003

Tempus Publishing Limited
The Mill, Brimscombe Port,
Stroud, Gloucestershire, GL5 2QG

British Library Cataloguing in Publication Data.
A catalogue record for this book is available from the British Library.

ISBN 0 7524 2879 9
Typesetting and origination by Tempus Publishing Limited
Printed in Great Britain by Midway Colour Print, Wiltshire

Contents

Preface 6

Foreword 7

Introduction 8

1. Merthyr Tydfil to Cardiff Docks in the South Wales Coalfield 9

2. Rhymney to Caerphilly in the South Wales Coalfield 36

3. Tredegar to Cwmfelinfach in the South Wales Coalfield 63

4. Beaufort to Crosskeys in the South Wales Coalfield 81

5. Blaenavon to Newport Docks in the South Wales Coalfield 110

Preface

This great coalfield is assumed by various authorities to be approximately one thousand square miles, which are distributed as follows: Breconshire, 74 square miles; Carmarthenshire, 228 square miles; Pembrokeshire, 76 square miles; Glamorganshire, 518 square miles and Monmouthshire, 104 square miles.

Of the above, nearly 846 square miles are exposed, about 153 square miles lie beneath the sea and about one square mile is covered by newer formations.

Like chapel on Sunday, safety lamps have lit the path for Welsh miners and protected them from harm. Light is the miner's most precious friend in the subterranean blackness and also his worst enemy. The risk of a naked flame igniting methane gas was a constant threat in the coalmines. Firedamp explosions killed many miners, like that on 14 October 1913 at the Universal Colliery, Senghenydd, when 439 men and boys died. A miner's life, his income and welfare are intricately bound to the light used in the pits.

Mining was the bedrock of South Wales, employing 230,000 men and producing 57 million tons of coal a year at its height in 1913. But the history of mining in Wales, which dates to pre-Roman times, also includes slate, lead, ironstone, copper and gold.

For centuries the only light came from tallow candles, which were pushed into nooks and crannies or secured on a miner's hat with a lump of clay. Some mine owners supplied the candles; others sold them at a profit. To save money miners made their own from waste animal fat. Candles posed many problems, the least of which was that the rats ate them. On the other hand they kept miner John Evans alive when he ate them while trapped underground for twelve days at a colliery near Wrexham in 1819. Candles were used until the end of the nineteenth century in non-coal mines but they could not be safely used in many coalmines where inflammable methane gas lurks in the deeper seams.

A series of tragic firedamp explosions led scientist Sir Humphrey Davy to invent the first oil-fired flame safety lamp in the Christmas of 1815, the best present a miner could have had.

The Davy lamp incorporated a wire grille around the flame, which absorbed the heat and reduced the chance of explosion. Improved design gave mines the Clanny lamp in 1839 which used a glass cylinder to protect the flame and the Marsaut lamp in 1871 in which a metal bonnet replaced the wire grille. From the 1850s onward the gassy steam coal collieries of South Wales turned increasingly to the safety lamp, prompted by disasters like the firedamp explosion at Llanerch Colliery, near Pontypool, Monmouthshire which killed 176 men and boys on 6 February 1890. An explosion four months earlier prompted the inspector of mines to urge the introduction of safe lamps. The colliery manager refused with tragic results. Despite the advantages of safety lamps, many Welsh miners had to buy their own and thus resisted their introduction.

By the 1930s electric hand lamps became common in the coalmines, to be replaced by helmet-mounted electric cap lamps following Nationalisation (Vesting Day) of the coal industry on 1 January 1947. The advent of battery-powered electric lights led to a dramatic drop in the number of cases of miners' nystagmus, an eye disease caused by working in poor light.

The familiar miner's safety lamp is still in use. At the start of every shift a colliery official goes around the mine with a safety lamp testing for methane, which creates a blue cap over the flame.

There were no greater mineral treasures than the riches of coal found in such abundance in the valleys of the South Wales Coalfield.

David Owen, Author and Archivist

Foreword

Having been invited to undertake the pleasant task of writing a short foreword to this, the latest of David Owen's books on the South Wales Collieries, it seems fitting to commence by mentioning the considerable amount of detailed research such an undertaking entails.

The result of this sustained effort is well represented in the following pages. Whereas this publication is in no way intended as a definitive work of reference, it is hoped that it will give pleasure to the lay reader, and wake many memories for the ex-miner.

Because considerable effort has been made with regard to the accuracy of captions and dates, together with other facts and figures, it may prove helpful to any reader interested in gathering historical coalfield data relative to what was once one of the greatest and most well known coalfields ever worked.

Harry Rogers
Coalmining Historian

Rhagair

Wedi cael fy ngwahodd i ysgrifennu rhagair i lyfr diweddaraf yr awdur David Owen ar Faes Glo De Cymru, mae'n addas i ddechrau trwy sôn am y maint sylweddol o ymchwil manwl mae gwaith o'r math yn golygu.

Mae canlyniad yr ymdrech yma i'w weld yn ddigon da ar y tudalennau a ganlyn. Er nad yw'r cyhoeddiad yma yn waith ac argraffiad terfynol yn nhermau cyfarwyddiadur, rwy'n gobeithio y bydd yn rhoi pleser mawr i'r darllenydd lleyg, ac yn dihuno atgofion cyn lowyr.

Oherwydd yr ymdrech manwl sydd wedi ei wneud i sicrhau cywirdeb yr is-deitlau a dyddiadau, ynghyd â'r ffeithiau a ffigyrau, fe fydd yn adnodd o gymorth i'r darllenydd sydd â diddordeb mewn casglu hanes y maes glo yng nghyd-destun un o'r ardaloedd glofaol mwyaf o rhan maint ac enwogrwydd a fu erioed.

Harry Rogers
Hanesydd Glofaol

Introduction

The present is heir to the past, the future captive to the present. Our heritage and history are important first and foremost because we live in a world shaped by our forebears, and to understand that heritage is to understand the forces that created our world. As such, I feel privileged to be invited to write this introduction to the latest publication, *South Wales Collieries – Volume Four*, by my old coal mining butty David Owen, who in this volume turns his attention to the Central Valleys of Merthyr Tydfil to the Eastern Valleys of Rhymney, Sirhowy, Ebbw and Afon Lwyd, Monmouthshire in the South Wales Coalfield.

As an ex-miner, who followed the trade of my father and grandfather before me, I feel these photographic memories of times past are part of that vital heritage, and the good and bad of those days alike should not be forgotten. They are an insight and a testament to the struggles of men and women who fought to provide for their families.

Alan Jones
Big Pit Tour Guide

Cyflwyniad

Olynydd y gorffennol yw'r presennol, ac mae'r dyfodol yng nghlwm yn y presennol. Mae ein treftadaeth a hanes yn bwysig oherwydd rydym yn byw mewn byd sydd wedi ei greu gan ein cyndeidiau, ac mae i ddeall y dreftadaeth yma yn arwain ni i ddeall y grymoedd a greodd y byd. Yn hyn o beth, rwy'n teimlo'n freintiedig i gael fy ngwahodd i baratoi cyflwyniad i'r cyhoeddiad diweddaraf am Weithfeydd Glo De Cymru, sef y bedwaredd gan fy hen gydweithiwr glofaol David Owen, sydd, yn y gyfrol yma yn troi ei sylw at Feysydd Glo De Cymru yng Nghymoedd Canolig Merthyr Tudful ac yng Nghymoedd Dwyreiniol Rhymni, Sirhywi, Ebwy a Afon Lwyd, Sir Fynwy.

Fel cyn löwr, a ddilynodd llwybr llafur ei dad a'i ddat-cu, rwy'n teimlo bod yr atgofion darluniadol yma o'r amser a fu yn rhan allweddol o'n treftadaeth, a ni ddylid anghofio y da a'r drwg o'r cyfnod yma. Maent yn gipolwg ac yn gofgolofn i'r ymdrech gan ddynion a menywod i ddarparu ar gyfer eu teuluoedd.

Alan Jones
Tywysydd Pwll Mawr

One

Merthyr Tydfil to Cardiff Docks in the South Wales Coalfield

Black Gold – Aur Du, the Story of Coal – Hanes Glo

We continue our journey in the South Wales Collieries series with Volume Four, from Merthyr Tydfil, Glamorganshire to Newport Docks, Monmouthshire. *South Wales Collieries Volume One* covers the Central Valleys of Rhondda Cynon Taf; *Volume Two*, the Lewis Merthyr Collieries, Trehafod Village and the Rhondda Heritage Park and *Volume Three*, from Ogwr to the Western Valleys of Afan, Neath, Dulais, Tawe, Aman, Loughor, Gwendraeth in Carmarthenshire and to Pembrokeshire in the far west of the South Wales Coalfield.

In the thirteenth century coal was being turned out at Llanvabon, and the monks of Neath and Margam were similarly employed and very likely used it for their temporal good. Another claim for ancient coal working is Swansea, the Norman lord William de Breos AD 1305, 'empowering the tenant to dig Pit coal at Byllywasted, without the hindrance of ourselves or heirs'.

In the centre of the old Norman castles of the thirteenth and fourteenth centuries stood the smithy, our earliest ironmaster, who used coal as well as charcoal in his labour at Morlais Castle.

The primitive mode of working in Elizabethan days was to drive a level and when they found the coal they worked holes, one for every digger, each miner working with candle light. Boys carried the coal in baskets on their backs to the entrance and work was carried out from 6.00 a.m. to 6.00 p.m. every day. These primitive operations did not extend much beyond the scratching of the surface and it was not until the closing decades of the eighteenth century that coal mining as a settled industry sprang into being in this area. This was due to the utilisation of coal for smelting iron, a development in which the first John Guest, of Dowlais – the founder of a family destined to play a large part in the industrial and commercial life of Glamorganshire – was a pioneer.

At the opening of the nineteenth century coal was worked on a considerable scale in the Merthyr and Dowlais areas, the output being in excess of the iron-smelting needs of the time and there are records of a proportion of that excess finding its way down to Cardiff and over mountain tracks into Herefordshire.

Towards the middle of the nineteenth century a few far-seeing men, outside the coterie of ironmasters, apprehended the immense values connoted on the new term, 'South Wales Smokeless Steam Coal', a term which, within a short period, was to attain and establish a standard of coal value the whole world over and men set to work as pioneers in beginnings so humble and difficulties so immense that no one who reads the story of their work can withhold a tribute to their courage and their faith.

British historians generally concur in the opinion that coal was well known here before the arrival of the Romans and was used by workers in brass. The Britons knew coal by the primitive name of -glo and the use of coal by the Romans in Wales was proved by an interesting discovery of a Roman villa near Caerleon.

The main factor in the tremendous growth of industrial areas like Merthyr Tydfil was that thousands of people moved into them to find work in the iron and coal industries. Most of the people who came to Merthyr moved fairly short distances from the rural parts of Glamorganshire, others came from a similar background in West Wales, with a few migrating from as far as the North Wales counties. Although skilled and unskilled workers also came from England and Ireland, the new population of Merthyr was overwhelmingly Welsh.

From the Welsh countryside the people of the new Merthyr Tydfil brought not only their language, but also their customs, traditions and religious beliefs. Sports such as foot-racing and fist fighting, festivals such as the Cwrw Bach (where people drank, sang and generally entertained themselves) and local eisteddfodau all flourished in the growing town of Merthyr. Public houses were at the centre of this cultural and social life. In 1847 it was noted that there were 200 pubs or beer shops in the village of Dowlais alone. Why would public houses and drinking have been so popular in a town like Merthyr? Although at first the churches and chapels seem not to have been against pubs (they often used them for meetings) the reputation which Merthyr got for drunkenness and fighting did eventually turn the religious bodies against them.

The mining of coal thrived as a result of the tremendous expansion of the iron industry. A great deal of coal was needed to smelt iron in the 1830s – 3½ tons for each ton of coal. It was this demand for coal from the ironworks which played the major part in the development of the coal industry in South Wales up to the 1860s. Most of the new collieries opened were in fact not really separate concerns at all for they were owned and developed by the iron companies.

The Welsh coal industry had been of outstanding importance throughout the past centuries. Well over 2,000 collieries have been thought to have operated in the old counties of Breconshire, Carmarthenshire, Pembrokeshire, Glamorganshire and Monmouthshire and well over 3,000 million tons of coal have been extracted from the Valleys of South Wales since intensive mining began nearly two centuries ago.

The industry was at the heart of the industrial revolution in Wales, participating in the development of new methods and technologies which were of world importance. At its peak in the early twentieth century, the South Wales Coalfield had the largest export of any in the world and dominated coal markets as far apart as the Mediterranean, South America, Africa and the Far East.

The ports of Cardiff and Barry became the leading ports in terms of tonnage handled anywhere in the world and nearly all of their trade was in coal.

The proportion of coal cut by machine increased from seventy-five per cent in 1947 to about eighty-six per cent in 1955 and could hardly be raised much higher. But both shot-firing and loading remained normal manual mining operations at this time.

Mechanisation by a combined cutting and loading process had, therefore, been a main development objective. This was generally called 'power-loading'. Only about five million tons of coal were power-loaded in Great Britain in 1947 and the only types of machines available were of limited application. The last eight years had seen the development of many new machines and about a dozen different types were available, while a number of others were in various stages of development. The Board were, therefore, in a position to apply power-loading on a much wider scale than ever before. In July 1955, they launched a far-reaching programme to establish power-loading units throughout all coalfields, wherever physical conditions permitted.

Until the 1950s the South Wales coal industry maintained a steady level of production and employment, but since that time there has been a continuing decline in the number of miners in employment. Most of the pits which have been closed still have coal left to mine but with natural gas, oil and coal available more cheaply from abroad, the demise of the industry has been inevitable.

Nowhere has the decline of the coal industry been more dramatic than in the South Wales Coalfield.

The first ambulance used by Dowlais Colliery, Merthyr Tydfil, Taff Vale, Glamorganshire in 1912. In 1886 a Royal Commission recommended the establishment of mines rescue stations, Crumlin and Aberaman in 1908, New Tredegar 1910. They did not become widespread until the Coal Mines Act of 1911 made them compulsory.

Right: Pen-Y-Darren No.1 and No.2 Colliery, near Dowlais, Taff Vale, Glamorganshire in 1885. On 9 July 1856 the accidents reports show that John Phillips was fatally injured. Pen-Y-Darren Colliery was owned in 1867 by Guest & Co. The cages were wound by the means of flat ropes. Pen-Y-Darren No.1 and No.2 Colliery was closed in 1893. The Glamorganshire Coalfield is assumed by various authorities to have been approximately 518 square miles. By the 1930s electric hand lamps became common in the coalmines, to be replaced by helmet-mounted electric cap lamps following Vesting Day, 1 January 1947. The advent of battery-powered electric lights led to a dramatic drop in cases of miners' nystagmus, an eye disease caused by working in poor light.

This poor quality image shows Cwmdu Drift Mine miners, Ynysfach, near Merthyr Tydfil, Taff Vale, Glamorganshire in 1900. In 1913 Cwmdu Drift was worked by the Crawshay Brothers and employed 200 miners. A drift is an entrance tunnel into a mine, which is driven through strata and coal seams into the required coal seam, normally at a downward inclination.

Remains of Lucy No.2 Drift, Blaencanaid, near Merthyr Tydfil, Taff Vale, Glamorganshire in 1967. Lucy Thomas Mine was known locally as the Pit-Y-Witw (Widows Pit). Lucy No.2 Drift was opened in 1905. Seams worked were the Four-Feet and Yard. Output prior to closure in 1955 was 19,000 tons. Lucy No.2 Drift was closed on 11 March 1955 by the National Coal Board.

Abercanaid Colliery, near Merthyr Tydfil, Taff Vale, Glamorganshire in 1896. The mine was owned by the Plymouth Iron Co. to feed their nearby iron works. No.1 shaft depth 50ft; No.2 shaft depth 393ft; No.3 shaft depth 180ft. Seams worked were the Lower Four-Feet and Nine-Feet. Abercanaid Colliery was abandoned in June 1902.

North Duffryn Colliery, Abercanaid, near Merthyr Tydfil, Taff Vale, Glamorganshire in 1905. The mine was originally owned and worked by Plymouth Iron Co. No.1 shaft depth 486ft; No.2 shaft depth 639ft. Seams worked were the Lower Four-Feet and Nine-Feet. North Duffryn Colliery was closed in 1969 by the NCB.

South Duffryn Colliery, Abercanaid, near Merthyr Tydfil, Taff Vale, Glamorganshire in the early 1900s. The mine was originally owned and worked by Hills Plymouth Co. Ltd. On 13 February 1856 the accidents reports show that David Thomas was fatally injured. South Duffryn Colliery was used for maintenance when acquired on Vesting Day, 1 January 1947 by the NCB.

Left: A shotsman's (shot-firer) detonator case (without a lock) and two detonators. A shotsman is a qualified official who fires shot holes in a district. The person must be qualified with a mines certificate, authorised and appointed in writing by the manager. Before firing shots, a shot-firer must test for gas, see that sentries are posted and that everyone has withdrawn from that zone. He must take proper shelter in a manhole and stone dust the whole neighbourhood of the shot which should be thoroughly covered with fresh incombustible dust before firing for a radius of 5yds and continuous with any road within 10yds of the shot.

Opposite, below: Gethin Colliery, Abercanaid, near Merthyr Tydfil, Taff Vale, Glamorganshire in 1893. The mine was sunk in 1852 by the Crawshay family of Cyfarthfa Castle. The colliery was the scene of an explosion on 19 February 1862, killing forty-seven men and boys, the youngest being eleven years old. A second explosion on 20 December 1865 killed thirty-four men and boys, the youngest also being eleven years old.

Right: Remains of Waunwyllt Level, near Merthyr Tydfil, Taff Vale, Glamorganshire in 1930. It was opened in 1818 and was closed prior to Vesting Day, 1 January 1947. The shotsman's equipment included an approved safety lamp (for all firedamp testing was imperative as such lamps have a re-lighting device and are adjustable by the user to admit air at the top of the lamp only when actually testing), and an approved electric cap lamp – both must be locked and he must examine them before entering the mine and a locked case with detonators only. He should check the number before leaving the magazine and keep the case locked on his person. The equipment also includes: a key to lock the detonator case and key to the exploder (both kept fastened to the person); a pricker for making a hole in the primer cartridge without opening it (this tool must be of copper or brass); a secure canister containing no more than 5lb of explosive of one kind only; an efficient and suitable exploder; a copper scraper to clean out the shot hole; a copper breakfinder to detect lateral and longitudinal breaks in the shot hole and a wooden rammer. The cable should be at least 25yds long and must not be allowed to touch power or lighting circuits and must be insulated and protected. A pocket knife for baring the wire of the cable ends was also included.

Gethin Pit colliers holing (undercutting) the Nine-Feet steam coal seam in 1893. The photograph was taken by Mathew Truran, one time manager of several Merthyr mines owned by the Crawshay family. The holing is being done some way up the section of coal (more usually seams were holed at the floor level of the seam). Gethin Colliery was abandoned on 21 July 1893.

Castle Colliery, Troedyrhiw, near Merthyr Tydfil, Taff Vale, Glamorganshire in 1901. The mine was opened in 1866 by the Crawshay Brothers (Cyfarthfa) Ltd. The Shaft depth was 999ft. In 1913 the manpower was 1,361 miners. Castle Colliery was abandoned on 23 January 1928.

A Gate Belt Conveyor

A gate belt conveyor is a mechanical device which gathers the coal as it leaves the coalface conveyor and transports it along the roadway to a loading point called a dump end where it is discharged into drams or mine cars.

Conveyor belting is made of layers, or plies, of woven cotton (cotton duck), which have been impregnated with rubber. These layers are enclosed in a rubber cover. It is the cotton duck that enables the belt to withstand the tension. The rubber protects the cotton from damage.

Belting is sent underground in 50yd rolls and when assembled on the conveyor structure, the lengths of belting are joined together by fasteners (joints). The gate conveyor belt width is usually 36in.

The whole of the conveyor, the drive and structure must be in one straight line. This, more than anything else, ensures that the belt will run centrally on all parts of the conveyor structure. The whole of the conveyor should be properly levelled, section by section, because if the sections are not level, the belt will tend to climb to one side of the structure. The belt joints should be made correctly; care being taken that the ends of each length of belting are cut square so that lengths of belting when joined together will run correctly on the conveyor structure. The belt should run centrally on the structure, the return drum and on the driving drums. Any defective joint, which is liable to break, should be repaired immediately.

The plough near the return drum and in the tension box must be maintained in proper working order so that the drum does not become clogged up with coal. The level of oil in the gearbox must be checked and maintained up to the oil level plug every day. The nipples on all drums at the drive head, jib head, loop take-up and tension end should be greased every week and grease, injected where necessary, into the rollers of the structure every month.

The transfer point is a place along a gate or trunk belt conveyor where the coal is transferred from one conveyor to another. It consists at some collieries of an intermediate structure on which is built a hopper or special chute to prevent spillage during transfer.

A certain amount of damage is caused to belting at transfer points by the blows from the lumps of coal or stone, which fall from one conveyor on to the next. This is especially marked if the coal falls on the belt immediately over a set of idlers. This damage may be reduced by covering the idlers with rubber.

The stage loader is a short belt or scraper chain conveyor in tandem with the gate conveyor. Coal from the face conveyor is conveyed by the stage loader to the gate belt. The advantage of using a stage loader is that the gate belt is kept away from the rough conditions that tend to exist near the face.

Anderson and Boyes (AB) Longwall Coal Cutting Machines

Coal cutting machines (cutters) are used underground to assist in breaking down the coal so that it can be loaded on to a conveyor. A coal cutter cuts into a portion of the seam making a slot along the coalface, after which the remaining part of the coal is cut down by the use of hand or pneumatic picks or is broken down by explosives.

The slot made by the cutting machine is known as the 'cut'. Normally the cut is 5in or 6in high and penetrates the seam to a depth of 4ft 6in, although cuts are sometimes made as shallow as 2ft and as deep as 7ft. If the cut is made at floor level, the seam is said to be 'undercut'. If halfway up the seam it is called a 'middle cut' and if at the top of the seam, near the roof, an 'overcut'. A coal cutter is made up of three main parts: the driving unit containing the motor, the haulage unit which carries the operator's controls and the gearhead, to which is attached the jib and cutter chain. The power of the driving unit may be 40, 50 or 60h.p. according to the type of machine.

With compressed air (blast) -driven machines the air is fed along the gate road in steel pipes through a main control valve to 2in diameter rubber hose and so to the machine where the air is controlled by another valve operated by a handle attached to the haulage end of the machine. The blast with its high pressure forces the teeth of both rotors apart, causing the rotors to revolve. This movement continues as each succeeding pair of teeth comes under the influence of the compressed air.

A flexible steel rope either ½in or ⅝in diameter and about 30yds long is used to pull the machine along the face. At one end of the rope there is an eye or loop for attaching to an anchor prop. The other end of the rope is wound around and clamped to a drum attached to the haulage unit.

The jib provides the framework round which the chain rotates. The cutter chain consists of a series of pick boxes joined together in various ways. The cutter picks are held on the pick boxes by a set screw. The pick boxes along the chain are set at different angles so that the picks are fanned out to cut on a series of lines. Always use a sharp pick in every box of the chain so that the point of each pick faces the direction in which the chain is to travel when cutting and always check the rope for any damage.

Women and Children on the Surface and Underground in Coal Mines

Shaftesbury's Coal Mines Act of 1839 ended female labour underground and the working of winding gear by children, although an Investigating Committee in 1842 still found instances of children aged four, five and six, employed underground. It was quite common for children aged seven and eight to work underground and children aged nine were expected to be employed in the Pit. It is almost beyond understanding that a civilised society should allow children of five years of age to work, not only in factories, but in coalmines and ironworks. Industrial Britain in the late eighteenth and early nineteenth centuries accepted such a labour force as a necessary part of producing much-needed raw materials as well as making the finished article.

Adult life for the children of a mining family in South Wales could begin at a very early age. Boys and girls as young as six years old worked underground in the early years of the coal industry. Many of them were employed as doorkeepers. Their job was to open and shut doors, which cut off sections of the workings underground and which helped to control the ventilation of the mine. In 1842, ten-year-old Elizabeth Williams who worked at a mine near Dowlais told a Government inquiry that she earned 2*d* a day as a doorkeeper (less than 1p in today's money). Every day she risked being involved in an accident with a full dram of coal. Some boys and girls as young as nine had the much more demanding job of dragging drams loaded with coal along the underground dramlines. If they stumbled and fell they risked being run over and crushed by the drams.

After 1842 it was illegal to employ children under ten years old underground. The age limit was later raised to twelve. But many children below this age continued to work at the coalface. Many boys followed their fathers down the pit as soon as they legally could, often on their twelfth birthdays.

Education for these children was a haphazard affair for much of the nineteenth century. The task of setting up schools was left to bodies such as the Anglican Church and the Nonconformist Chapels. Sunday schools provided much of the education ordinary people received.

An Act of 1870 allowed School Boards to be set up. These created elementary schools providing a basic education for children from about the age of four to twelve.

In 1889 the Welsh Intermediate Education Act allowed most local authorities to set up 'county schools' with the power to educate children until they were eighteen, but most children still left school at the age of twelve.

Merthyr Vale Colliery, Aberfan, Taff Vale, Glamorganshire in 1900. Merthyr Vale Colliery was the last venture of John Nixon, who, with his partners, commenced sinking on 23 August 1869 and proved the Four-Feet seam on 1 January 1875. The first two drams of coal were raised on 4 December 1875. According to Charles Wilkins writing in 1888, 'The colliery was known as the premier colliery in South Wales and second only in depth and magnitude of scientific appliances to Treharris'. The two shafts were sunk about 255ft apart; the downcast and winding shaft was 16ft in diameter. The upcast, also a winding shaft of 16ft in diameter, was closed at the top by two wrought iron hoods, arranged so as to be lifted by the cage on its ascent.

The Merthyr Vale Colliery winding engine on the downcast shaft was in fact one of the very few engines which had been built as a marine engine for a large steamship. It was built by Maudslay, Son & Field of London; it had two 83in horizontal cylinders with a 4ft stroke, two piston rods worked through the cylinder top to a crosshead, working directly to the shaft of a spiro-cylindrical drum which was 12ft diameter at the first lift and 24ft at the terminal lift. This drum was 16ft wide at the middle and weighed 60 tons. Even at the low pressure of 30psi. this was probably the most powerful engine for winding in the country in the 1890s. The cage carried two drams on one deck and was raised in thirty-eight seconds. In 1891 the colliery was ventilated by a Waddle fan, 40ft in diameter, which produced 230,000 cubic feet of air per minute.

For many years the colliery was owned by Nixons Navigation Steam Coal Co., but the concern was later to be known as Llewellyn (Nixon) Ltd. In the 1930s the colliery became part of the largest coal group in the UK, the Powell Duffryn Associated Collieries Ltd (PDs) known to many as the 'Poverty and Dole' Group who worked the colliery until Nationalisation. Among the colliery undertakings in the South Wales Coalfield was that of the Powell Duffryn Empire; in fact it may be said that it was one of the largest, most modern and best equipped in the world at the time.

Prior to 1960, Merthyr Vale operated as two pits, the No.1 shaft mining the Four-Feet seam and the No.2 working the Five-Feet/Gellideg. A £2 million reorganisation in the early 1960s integrated the two operations, electrified the winding and increased underground conveyor capacity. Coupled with a new coal preparation plant (washery) and rail wagon-loading facilities on the surface, the scheme geared the pit to take its place amongst the coalfield's major production units.

Merthyr Vale Colliery, underground multi-jibbed coal cutter in 1979. The workings were beneath an area of around four square miles and concentrated on the Seven-Feet seam for supplies of top-quality dry steam coals for the manufacture of patented domestic smokeless fuels. There were more than ten miles of underground roadways within the mining programme and around three miles of high-speed belt conveyors in daily use. The 'take' was bounded by 90ft Werfa (Kilkenny) Fault in the west and by the workings of the nearby Taff Merthyr and the Bedlinog Colliery (which ceased production in 1924 and closed on 26 March 1956).

Following the publication of *Plan for Coal* in 1973, a massive capital investment programme was launched by the National Coal Board (NCB), aimed at securing the nation's vital energy supplies well into the next century. For South Wales alone, over £100 million was earmarked for major modernisation projects to streamline long-life pits, to locate and log new coal reserves, to create new mines and to link established neighbouring pits into single, high-production units – Britain's guarantees in a fuel-hungry future! This colliery was one of those involved in the programme.

In 1974, Merthyr Vale was in the vanguard of the NCB/NUM joint production drive and quickly set the standard with a new coalfield productivity record of 45cwt per shift for every man on the colliery books. A year later, a coalfield campaign for 'tidy pit surfaces' brought the pit not only an NCB award, but a certificate of commendation from the European Architectural Heritage Year Business and Industry Panel.

In 1976 with a manpower of 621, Merthyr Vale Colliery produced an annual saleable output of 240,000 tons; it produced an average weekly saleable output of 5,000 tons; average output per man/shift at the coalface 4 tons 4cwt; average output per man/shift overall 1 ton 9cwt; deepest working level 2,540ft; number of coal faces that were working 2; No.1 shaft depth 1,627ft, diameter 16ft; No.2 shaft depth 1,607ft, diameter 16ft; manwinding capacity per cage wind 24; coal winding capacity per cage wind 6.5 tons; winding engines horsepower 1,450; stocking capacity on pit surface 87,000 tons; average weekly washery throughput 7,000 tons; types of coal dry steam; markets manufactured smokeless fuels; fan capacity, cubic feet per minute 158,000; average maximum demand of electrical power 4,500kW; total capital value of plant and machinery in use was £904,886; estimated workable coal reserves 7.1 million tons. Merthyr Vale Colliery was closed in August 1989 by British Coal.

'Friday, 21 October, 1966 – A Date Which Will Live In Infamy' – The Valley Laid Silent

The last major disaster in the South Wales Coalfield and the one which more than any other, broke the heart of a nation, happened at Aberfan, Taff Vale, Glamorganshire and became the worst mining catastrophe in the history of industrial Britain, with its appalling toll of 144, including 114 school children, 2 children not of school age and 28 adults, 6 of them teachers. This date must never be forgotten. What made the tragedy at Aberfan all the more poignant was that, more than any other in the South Wales coal industry, it was avoidable.

Friday 21 October 1966 is one date in the history of the South Wales Coalfield, which for sheer horror surpasses all others. On that date the anonymity of a small Welsh mining community was lost forever. Front pages of newspapers the world over carried headlines of shock, disbelief and sympathy.

John Nixon began sinking Merthyr Vale Colliery with his partners on 23 August 1869 but before his arrival the tiny community of Aberfan consisted only of two cottages and an inn overlooking a green and pleasant land. Yet, following Nixon's sinking of the colliery, the landscape had been transformed by the growth of ugly slag heaps (tips) which destroyed the natural beauty of the valley.

Such was the horror that unfolded on a cold, damp, grey October morning that it is hardly surprising that, even today, some villagers still cannot talk about what happened. Yet, it is impossible not to remember the children of Aberfan alongside all those other thousands who gave their lives for coal. They had been looking forward to their half-term holiday that morning, arriving at Pantglas Junior School for what was only due to have been a half day. But high above them, among the tips that towered over the village, the tragedy was swiftly unfolding.

The No.7 Pantglas Tip had been shrouded in controversy since its origins five years earlier. There was already evidence that previous tips built there had been constructed on boggy ground. Years later it was revealed that the No.7 Tip itself had an underground mountain stream bubbling away directly underneath it. But despite protest, work went ahead on the new tip in 1961. The following year, the NCB began using it to deposit 'tailings', minute particles of coal and ash which, when wet, took on a consistency similar to quicksand.

Friday 21 October had dawned cold, misty and overcast, three inches of rain having fallen in the previous week. Engineers at work on the No.7 Tip had arrived there at 7.30 a.m. to find a 30ft crater in the centre. Supervisors were summoned to survey the situation but, almost at the moment they arrived at 9.00 a.m. a huge wave burst from the tip, carrying with it hundreds of tons of debris. The black torrent raced towards the still hidden village, engulfed hillside cottages and uprooted trees which stood in its path. The slithering mass formed a wall some 30ft high, carrying beneath it the debris of the demolished houses that had crumpled in its relentless advance. It sped towards Pantglas Junior School, crushing the walls and roof of the school without faltering. All emergency services had been alerted. On hearing the news, Merthyr Vale miners rushed to the stricken village, where residents had begun tearing at the mountain of rubble with their bare hands. It was the colliers, most still black from their work underground, who gave their expertise and provided invaluable assistance by directing the essential digging operations. Side by side miners, frantic parents and alarmed onlookers dug in a desperate attempt to uncover those buried beneath the falls of rubble. The news of the disaster reached far and wide, people began to flock to the scene. Most came to offer help, others simply to stare. The first message of Royal sympathy came from the Prince of Wales. A disaster fund was set up by the Mayor of Merthyr Tydfil and thousands of messages of condolence were received.

Thankfully, the tribunal which investigated the disaster agreed where the blame should be laid. The tribunal chairman said the NCB's liability was 'incontestable'.

A sad reminder of the true price of coal, and it must never happen again.

Aberfan Memorial Garden entrance in 2003. The memorial reads:

PANTGLAS JUNIOR SCHOOL MEMORIAL GARDEN DEDICATED TO 116 CHILDREN AND 28 ADULTS WHO LOST THEIR LIVES 21 OCTOBER 1966 TO THOSE WE LOVE AND MISS VERY MUCH I'R RHAI A GARWN AC Y GALARWYN O'U COLLI

Above: Aberfan in 2003. Tips and abandoned workings now lie hidden beneath the gentle slopes of reclamation; the industrial landscape of Wales has changed to being green and lush and nature has returned as a result. We retain our sense of pride and traditional heritage but combine it with a new optimism.

Opposite, below: Bedlinog Colliery, Bedlinog in 1895. The mine was opened in 1881 and was owned by Guest Keen and Nettlefolds Ltd. No.1 shaft depth 1,746ft (1881); No.2 shaft depth 1,740ft (1883). Seam worked was the Brithdir. In 1909 with a manpower of 2,242 it produced an annual output of around 469,000 tons. Bedlinog Colliery was closed on 31 March 1956 by the NCB.

Above: Colly Level, Bedlinog, Taff Vale, Glamorganshire in 1912. In the 1870s Colly Level was worked by the Dowlais Iron Co. A level is a tunnel driven horizontally or on a slight gradient to connect underground workings with the surface. Colly Level was abandoned on 31 August 1961. In 1794 the Glamorganshire Canal was completed between Merthyr and Cardiff, the distance covered twenty-four miles, forty-nine locks, and a rise in level of 543ft to Merthyr.

Nantwen No.8 Level, Taff Vale, Glamorganshire in 1965. In 1966 Nantwen No.8 Level was licensed and owned by F.H. Price, c/o 58/59 Lower Thomas Street, Merthyr Tydfil and worked the Six-Feet seam under the inspectorate of the South Western Division. Nantwen Level was abandoned on 18 January 1966.

High Street Level, Taff Vale, Glamorganshire in 1973. The mine worked the Brithdir seam. High Street Level was abandoned in the 1970s. In 1830 the Steamship Companies chose Welsh steam coal for its efficiency. In 1839 the West Bute Docks at Cardiff opened, bringing with it a new era of development to the South Wales Coalfield.

Ropesmith Mark Owen at Taff Merthyr Colliery, near Treharris, Taff Bargoed Valley, Glamorganshire, 20 March 1991. Coal mining dominated the Taff Bargoed Valley through the nineteenth and early twentieth centuries and the closure of Trelewis Drift in August 1989, Deep Navigation Colliery in March 1991 and Taff Merthyr Colliery with the final shift worked on Friday 11 June 1993, brought to an end a proud and historic chapter in the history of the valley. Loss of the mining industry stripped the communities of an economic means of existing, as well as plunging the physical heart of the community into dereliction. The Taff Bargoed Valley lost 2,000 jobs and £6 million from the local economy as a result of pit closures. The landscape was left derelict and scarred having a knock-on effect on the social well-being of those living and working locally. Local Authority engineers then began to develop land reclamation proposals which were later to develop into the multi-faceted Taff Bargoed Community Park, Climbing Centre complex and a water remediation scheme now in place. The Coal Authority constructed a minewater treatment system at the site of the former Taff Merthyr Colliery. The largest minewater wetland scheme in the country is the key feature of the Taff Bargoed Millennium Park, a major reclamation project undertaken by Merthyr Tydfil County Borough Council and Groundwork Merthyr & Rhondda Cynon Taf.

Taff Merthyr Colliery was sunk in 1926 by Taff Merthyr Steam Coal Co. to a depth of 1,903ft and was set up by the Powell Duffryn and Ocean Co.'s. Seams worked were the Four-Feet, Six-Feet, Seven-Feet, Five-Feet, Yard, Bute and the Gellideg. The Colliery encountered significant water problems during its operational life, after which time the two shafts were filled and capped and the associated colliery buildings were demolished. Water continued to fill the shafts and surrounding workings and in November 1994, minewater flows of up to 120l per second, containing around 20 milligrams per litre of dissolved iron salts, began to flow from the two shafts via culverts close to the surface. Although a small temporary treatment system was installed, the bulk of the iron 'ochre' flowed to the Bargoed Taff with consequential orange staining and downgrading of the river for many kilometres downstream. With the promotion of the Taff Bargoed Millennium Park, there was clearly a need to phase the work associated with the remediation of the discharge with that being undertaken to create the Community Park, on land incorporating the former Taff Merthyr Colliery, Deep Navigation Colliery and Trelewis Drift Mine.

Taff Merthyr Colliery underground hauliers at work in the 1930s. The Taff Merthyr Project has been designed to ensure that minewater discharged to the Bargoed Taff, which feeds the Community Park lakes, remains clean. The minewater treatment scheme has been constructed by the Coal Authority on the derelict former colliery site and extends to some seven hectares, containing over three hectares of wetland treatment. Treatment consists of four large settlement lagoons, sixteen reed beds, a pumping/aeration system, access roads, cycle track and extensive landscaping. Pumping is necessary to ensure that the whole site is utilised for treatment. Minewater flows are split and controlled into four treatment 'legs'. Standby pumps and telemetry systems are incorporated to ensure that the minewater is treated at all times.

The scheme successfully treats the variable minewater flows from the old colliery workings, ensuring that the cleaned discharge to the Bargoed Taff contains less than 1 milligramme of iron per litre of water. This successful treatment operation prevents approximately 72 tons per year of iron sediment entering the Bargoed Taff.

The scheme has the added benefit of restoring a previously derelict site, and forms an integral part of the wider Community Park. Cycle tracks provide public access links to the valley and extensive landscaping enhances the visual aspect of the site. The reed beds themselves are forming a valuable wildlife habitat for a variety of species that visit the ponds and surrounding areas, and this is contributing to the biodiversity of the area.

From 1993 onwards the ravaged landscape of three former colliery sites in the Taff Bargoed Valley has been transformed by a community based scheme with sustainable development principles at its heart.

The tips were transformed and reshaped as the initial reclamation requirements were adapted through consultation and debate. The original designs were altered to include a series of lakes and weirs for fishing and canoeing. On the Deep Navigation Colliery site a pavilion has been constructed along with new sports facilities and six new bridges now span the site's watercourses. Local schoolchildren have taken part in educational projects such as 'Trees of Time and Place' and 'Green IT'. Community arts, video projects and training courses have been run to give confidence, boost creativity and broaden horizons. This has added value to the project, helped form further partnerships and helped ensure that the park's development remains needs-driven.

Taff Merthyr Colliery underground horsehead bars in the supply road in 1966. This type of horsehead bars are supporting the ripping lip and ventilating the face beneath the rippings. Whenever there was a presence of methane gas a blast (compressed air) hose was put into the ends of the hollow horsehead bars. The blast would travel through the horseheads and dilute the gas to a safe mixture for work to continue in safety. The ripping lip is the removal of stone from above the seam to create a higher heading at the entry to the coalface in preparation to erect rings (steel arches) which are then lagged with timber. The road was ripped on the afternoon and night shifts. A conveyor in the gate road carries the coal outbye. The supply road supplies the face and district with timber, props, rings, rails and sleepers etc. Taff Merthyr Colliery was closed on Friday 11 June 1993 by British Coal.

Trelewis Drift Mine, Trelewis, Taff Bargoed Valley, Glamorganshire in 1953. Trelewis Drift Mine was started in 1952 by the NCB and coal production began in the Brithdir seam. In 1955 with a manpower of 88 it produced an annual saleable output of 29,000 tons. Trelewis Drift Mine was closed in August 1989 by British Coal.

Above left: Deep Navigation Colliery pit bank in 1915. The bank is the surface land surrounding the mouth of a shaft and the surface of a shaft and at a level from which the pit cages are loaded or unloaded.

Above right: Deep Navigation Colliery boring rig in 1975. Exploratory holes were bored to define the coal seams, minewater, methane gas etc. in the vicinity.

For decades, Deep Navigation had been one of the most productive and profitable mines in the South Wales Coalfield.

Within the colliery premises was housed the Treharris Boys Club, well known in the locality particularly for its boxing record. The miners had traditional links with the club and many of the boys ultimately worked at Deep Navigation Colliery.

Deep Navigation Colliery Coal Seams:

Depth Feet	*Standard Name*	*Local Name*
2,070	Four-Feet	Four-Feet
2,130	Six-Feet	Six-Feet
2,240	Upper Nine-Feet	Upper Nine-Feet
2,275	Lower Nine-Feet	Lower Nine-Feet
2,385	Seven-Feet	Seven-Feet
2,425	Five-Feet/Gellideg	Five-Feet/Gellideg

Two years after the centenary celebrations the colliery recorded the best ever profit of £2.6 million and in the second week of May 1981 the 780 men hit a productivity record of 3 tons 1cwt per man shift. The total saleable output for the week was 9,595 tons.

In 1985 the colliery was equipped with one of the most expensive and massive powered supports which weighed up to 18 tons and the face when in production could cut up to 1,500 tons of coal in one day. During the first week in March the high-technology face cut a saleable output of 9,000 tons, with a coal face productivity reaching 28 tons per man shift. By March 1986 the colliery production reached half a million tons of saleable coal for the year. Deep Navigation Colliery was closed in March 1991 by British Coal.

A painting showing Cefn Glas Colliery in a scene at the railway junction in Quakers Yard, Taff Vale, Glamorganshire. Upper right in the photograph is a representation of Cefn Glas Colliery alongside the Glamorganshire Canal. The mine was sunk in 1863 and had a chequered history of closures. Cefn Glas Colliery was abandoned in February 1894.

Treforest School of Mines, Taff Vale, Glamorganshire in 1910. Seated far right in the photograph is Hubert Clarke who later became head surveyor at Cilely Colliery, Tonyrefail, Ely Valley, Glamorganshire. In 1841 the Taff Vale Railway opened between Merthyr and Cardiff to ease the congestion affecting the canal systems and Cardiff became the great coal exporting centre of South Wales.

Nantgarw Colliery locomotive skid patch in September 1985. Alan Jeynes (loco fitter) and David 'Dai Diesel' Davies (driver instructor) are in the cab of the loco. The rails of the skid patch were set at an incline and were greased prior to setting off. All types of loco control, including progressive breaking, were administered during the course.

Nantgarw Colliery miners in November 1986. The photograph includes: Ivor Nicholls (blacksmith), Trevor Williams (mine car repairer), Joe Davies (assistant engineer), Noel Stevens (blacksmith), Dennis Arthur (blacksmith), Keith Harris (turner fitter), Owen Rees (welder).

Nantgarw Colliery pit wheel, 7 August 1999. In 1976 with a manpower of 587 it produced an annual saleable output of 143,306 tons; an average weekly saleable output of 3,936 tons; average output per man/shift at the coalface 5 tons 1cwt; average output per man/shift overall 1 ton; deepest working level 2,247ft below the surface and 2,000ft below sea level; number of coal faces that were working 3; No.1 shaft depth 647yds, diameter 19ft; No.2 shaft depth 605yds, diameter 19ft; manwinding capacity per cage wind 26; coal winding capacity per cage wind 6 tons; winding engines horsepower 1,450; stocking capacity on pit surface 600 tons bunker en route to washery; average weekly washery throughput 4,000 tons; types of coal, high quality coking; markets coke ovens and steel industry; fan capacity, cubic feet per minute 250,000; average maximum demand of electrical power 3,687kW; total capital value of plant and machinery in use was £1,043 million; estimated workable coal reserves 9.2 million tons.

In 1977 losses were over of £1 million but by May 1978 the loss was transformed to a profit of over £40,000 per month, for a chainless haulage system, and an Anderson Strathclyde ranging drum shearer had raised output to over 5,000 tons per week. At this time the F44 coalface produced a record 334cwts per manshift, which was amongst the best in the South Wales Coalfield and the best ever recorded by Nantgarw/Windsor.

A 1978 NCB press release stated that 'Nantgarw/Windsor is one of the coalfield's premier collieries, linked by a £1.75 million project. The colliery will exploit over 9 million tons of coking coal, which will keep the mine in business well into the next century'.

In 1985 Nantgarw incurred losses of £5 million, and in 1986 a further £4 million was lost in 6 months. British Coal said that output results from two districts affected by worsening and uncertain geological conditions had continued to decline. The area director said that he could see no justification for continuing mining operations at Nantgarw, because of the colliery's uncertain geology.

Depth Feet	*Standard Name*	*Local Name*
2,007ft	Two-Feet-Nine	(Un named)
2,016ft	Upper Four-Feet	(Rider)
2,064ft	Four-Feet	(Two-Feet-Nine)
2,100ft	Six-Feet	(Four-Feet)
2,145ft	Upper Nine-Feet	(Six-Feet)
2,169ft	Lower Nine-Feet	(Lower Six-Feet)
2,259ft	Bute	(Red Vein)
2,280ft	Amman Rider/Yard/Seven-Feet	(Nine-Feet)

Nantgarw/Windsor Colliery was closed on 7 November 1986 by British Coal.

Remains of Groeswen Level, also known as Cattie Ddu (Black Katie), Nantgarw, Taff Vale, Glamorganshire in 1969. Seams worked were the Brass and Black Vein. Groeswen Level was closed in 1950 by J.E. Osborne. In 1866 work began on Roath Basin at Cardiff Docks.

Above: Remains of Lefel-Y-Cwcw (The Cuckoo Level), Garth Mountain, near Taffs Well, Taff Vale, Glamorganshire in 1970 (photograph by Gordon Hayward). This type and size of mine was known by some old miners as 'Rat Holes' or 'Tobacco Box' mines. Seam worked was the Penygroes. Lefel-Y-Cwcw was abandoned in the 1950s.

Opposite below: Excelsior Wire Rope Works, Llandaff, Cardiff in 1907. Excelsior Wire Rope Works manufactured wire ropes for cage guides, underground and surface haulage engines including winding ropes for single rope drum winding and multi-rope installations at South Wales Collieries.

Above: New Rockwood Drift Mine, Taffs Well, Taff Vale, Glamorganshire in 1965. The drift mine was opened and worked in 1936 by New Rockwood Colliery Ltd until Vesting Day, 1 January 1947. Seams worked were the Brass and Black Vein. In 1961 with a manpower of 160 it produced an annual saleable output of 48,000 tons. New Rockwood Drift Mine was closed on 25 May 1963 by the NCB.

Above: National Union of Mineworkers Gala, Cardiff City, Glamorganshire in 1959. The photograph includes: Glyn Williams (President), W.H. Crew (South Wales Area General Secretary 1951-1958), D.D. Evans (South Wales Area General Secretary 1958-1963), Ness Edwards, Ben Morris MP, Tom Mantle, Joe Hughes, William Whitehead (President 1959-1966), L.R. James, Aneurin Bevan, Tommy McGee MP, Abe Moffat, Wendall Jones, Jack Jones, W.R. Jenkins, Dai Francis (South Wales Area General Secretary 1963-1976).

Below: Cardiff Docks, Glamorganshire in the 1930s. The speed with which the coal industry expanded in South Wales is reflected in the growth of Cardiff during the nineteenth century. It grew from a village with just 1,870 inhabitants in 1801 into Wales's largest town with 164,333 people a century later. By then it was on its way to becoming the greatest coal exporting port in the world. It was officially declared a city in 1905 and the capital city of Wales in 1955. Cardiff Docks exported 10,278,963 tons of coal in 1914.

In Everlasting Memory to the Miners who lost their lives from Merthyr Tydfil to Cardiff

Date	Mine	Lives Lost
2 July 1852	Gethin, Merthyr Tydfil	Sinker Elias Roberts (18) killed by falling down the pit.
1 October 1852	Rhas Las, Dowlais	Colliers John Davies (29), Jacob Griffiths (34) and Humphry Edwards (24) killed when the guide-chain broke and the carriage struck the side of the shaft.
10 December 1852	Dan-Y-Deri, Merthyr Vale	Collier Thomas Parry (24) killed by fall of roof.
23 March 1853	Abercanaid, near Merthyr	Collier William Jenkins (14) killed when he was thrown off the cage by a mandrill.
22 January 1872	Merthyr Vale	Sinker T. Chaple (36) killed by shot going off at bottom of sinking shaft.
2 September 1872	Pentyrch, near Cardiff	Collier Isaac Morgan (28) killed by a fall of stone.
6 December 1875	Llan, Pentyrch	Sixteen colliers killed by an explosion of gas.
4 February 1886	Deep Navigation, Treharris	Collier Dennis Manning (39) killed by a stone falling on his head.
12 June 1890	Nantgarw, Taffs Well	Colliery Edward Davies (21) killed by a fall of a bell-shaped stone from roof close to heading face, which was not timbered. Pillar and Stall method was in use.
9 September 1890	Mountain Level, Merthyr	Haulier William Jones (21) killed by a fall of roof on a level when passing with a horse and dram. Two pairs of double timbers, 2ft apart and some lagging were thrust out by the fall, which went up 9ft to the surface.
23 January 1891	Plymouth, Merthyr	Collier John Jones (20) killed by a fall of roof 12ft x 4ft x 1ft thick at the face of the heading in the Five-Feet seam while cutting bottom. Two side props and the top of a cog on the opposite side were crushed out by the fall, which burst out of the solid rock roof. The road was 10ft wide.
10 February 1891	Pen-Y-Darren, Merthyr	Rider David Thomas (32) killed while following a journey of four drams up an incline, the hitching-plate of the first dram broke, then the bar-hook attached to the last dram gave way and the drams ran back upon him, crushing him to death.
30 April 1891	Merthyr Vale	Master Haulier Richard Hughes (30) killed while riding between drams of rubbish on engine-plane; it is supposed that a dram left the rails, struck out an arm, and that the collar in falling knocked him off under the drams.

Once again a sad reminder of the true price of coal.

Two

Rhymney to Caerphilly in the South Wales Coalfield

Candles were used by miners for artificial light underground. Some were made of animal fat and were occasionally eaten by rats and mice; one miner was kept alive by eating the candles when he was trapped in a pit for twelve days. Regular catastrophic explosions spelled the end for naked flames, however, although there was resistance to the safety lamp. The problem was that until the end of the nineteenth century, the safety lamp gave far poorer light than a candle. Colliers were paid by their output, so less light meant less coal. They also had to pay for their own lamps at first. Legislation in 1911 outlawed naked flames in deep mines, where gas was more likely, although miners continued to use them until the 1930s.

Like chapel on Sunday, safety lamps have lit the path for Welsh miners and protected them from harm. Light is the miner's most precious friend in the subterranean blackness and also his worst enemy. The risk of a naked flame igniting methane gas was a constant threat in the coal mines. Many miners were killed by firedamp explosions, like that at Universal Colliery, Senghenydd when 439 men and boys were killed. A miner's life, his income and welfare are intricately bound to the light used in the pits.

Mining was the bedrock of South Wales, employing 230,000 men producing 57 million tons of coal a year at its height in 1913. But the history of mining in Wales, which dates to pre-Roman times, also includes slate, lead, ironstone, copper and gold.

For centuries the only light came from tallow candles which were pushed into nooks and crannies or secured on a miner's hat with a lump of clay. Some mine owners supplied the candles; others sold them at a profit.

A series of tragic firedamp explosions led scientist Sir Humphry Davy to invent the first oil-fired flame safety lamp in the Christmas of 1815, the best present a miner could have. The Davy lamp incorporated a wire grille around the flame which absorbed the heat and reduced the chance of explosion. Improved design gave mines the Clanny lamp (Dr Clanny) in 1839 which used a glass cylinder to protect the flame and the Marsaut lamp (M.J.B. Marsaut) in 1871 in which a metal bonnet replaced the wire grille. From the 1850s onward the gassy steam coal collieries of South Wales turned increasingly to the safety lamp, prompted by disasters like the firedamp explosion at Llanerch Colliery in Monmouthshire which killed 176 men and boys on 6 February 1890. An explosion four months earlier prompted the inspector of mines to urge the introduction of safe lamps. The colliery manager refused with tragic results.

Despite the advantages of safety lamps, many Welsh miners had to buy their own and thus resisted their introduction.

Pwllaca Pit, Rhymney, Rhymney Valley, Glamorganshire in 1906. The mine was opened in 1879 by the Rhymney Iron Works Co. The shaft was oval with a depth of 399ft. The name was changed to McLaren No.2 Colliery by the Tredegar Iron & Coal Co. in 1899 when it became the upcast shaft for McLaren No.1 Colliery which was closed on 18 July 1958 by the NCB.

Fochrhiw No.1 and No.2 Colliery, Rhymney Valley, Glamorganshire in 1900. Fochrhiw No.1 and No.2 Colliery was opened in the Mid 1800s by Guest & Co. On 14 June 1867 the accidents reports show that thirteen-year-old collier George Bowen was fatally injured. Fochrhiw No.1 and No.2 Colliery was abandoned on 30 June 1924.

Ffynon Duon No.1 Level, Fochrhiw, Rhymney Valley, Glamorganshire in 1970. In the photograph is a pit pony, a miner's friend and companion. In 1930 there were approximately 11,500 horses employed underground in the South Wales Coalfield. Several pit ponies had over eight years service underground. They were treated like family pets by the hauliers and not to be left out, some horses enjoyed a chew of tobacco to keep the dust at bay.

Ffynon Duon No.1 Level in 1974. Ffynon Duon No.1 Level was opened 1925. In 1967 the owner was John P. Llewellyn, Glen View, Pentwyn, Fochrhiw and worked the Brithdir seam under the inspectorate of the former South Western Division. Ffynon Duon No.1 Level was abandoned in the 1970s.

A Miners' fight led to a General Strike on 3 May 1926

Between 1914 and 1918 the First World War meant Welsh coal was kept in high demand because it was used to fuel the steam engines that drove the ships of the Royal Navy and the Merchant Navy. However so many miners volunteered for the Army that those left struggled to produce the amount of coal the ships needed. Those who remained in the mines demanded higher wages for their work and they threatened to strike if their demands were not met. The Government feared that strikes might cripple the Navy by robbing it of the coal it needed, so it took over the running of the mines from the coal owners. Under the Government's management the miners' wages rose.

After the war, control of the mines returned to private businessmen who said that the industry could not afford to continue to pay such high wages. They demanded wage levels should be cut; the miners refused and in 1921 found themselves locked out of the mines and jobless for three months. In 1925, the owners decided they needed more wage cuts and once again the miners protested. The Government stepped in with subsidies to keep wages at their current levels while an enquiry known as a Royal Commission examined the industry. It reported in 1926 and said the industry should be reorganised. Some wage cuts would eventually be needed, it said. The owners demanded immediate cuts. Once again the miners refused and once again they found they had been locked out of their mines.

Communal kitchens set up for the miners' families frequently served up lots of high quality food; by the end of it those families were surviving on dry bread and bully (corned) beef.

Fellow workers decided to support the miners and a General Strike began on 3 May 1926, which threatened to bring the whole of Britain to a standstill. A sense of alarm spread through that section of society, which supported the Government and the coal owners. They felt the country stood on the brink of a revolution.

Many of them volunteered to help run threatened services such as buses and trains. Troops were called in to ensure these services were not under threat from disruptive action by strikers. But before the strike could cause too much damage it was called off. The Trades Union Congress, which had organised the strike, wrongly argued that the Government was now prepared to negotiate a settlement to the dispute. The strike had lasted just nine days.

The miners felt betrayed by their fellow workers and refused to go back to work on the terms of the coal owners. They fought on, despite receiving no unemployment pay. The only financial aid for the miners came from occasional strike pay provided by their unions and relief money provided by those whose job it was to help the desperately poor, the Poor Law Guardians. The money the strikers received was never enough to live on. Local authorities could also provide help, largely in the form of communal meals.

Boot centres were established which repaired the shoes of miners and their children. A repair centre at Ynysybwl, near Pontypridd, repaired almost 800 pairs of children's boots and 350 pairs of adult footwear during the seven month lock-out.

Miners spent much of their time raking over tips to collect the pieces of coal they needed to heat their homes. But as the lock-out went on the miners' plight became more desperate. Some men did try to return to work.

Violent clashes broke out between striking miners and police trying to protect the 'blacklegs', as those returning to work were called.

Despite the sense of unity among the miners and their spirited determination not to give in, they were eventually forced to accept the coal owners' terms and wage levels. They returned to work on 1 December 1926.

What helped to save the miners' cause was their very strong sense of community spirit. It was expressed in a wide variety of ways. They organised carnivals and parades, sports events and concerts. Comic jazz bands were formed.

Above: South Wales Coalfield hauliers in 1931. Pit ponies were well cared for by the hauliers; stables generally were warm, dry and comfortable and many were lit by electricity. Moss litter or sawdust was always provided for bedding. Plenty of good food and clean water was at hand, both at the stables and while at work, and a sufficient supply of medicine and dressing was readily available.

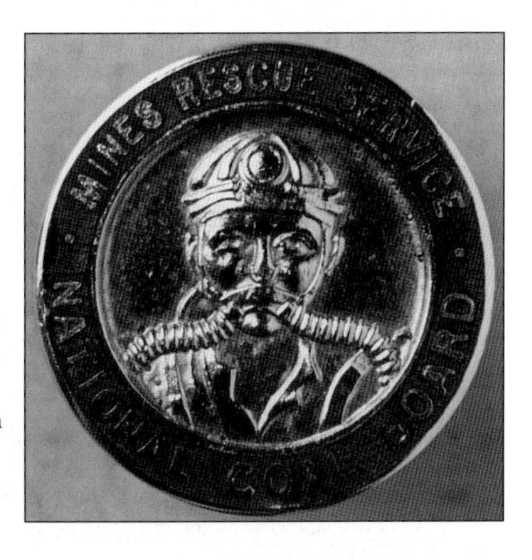

Right: Kerrigan Jones' NCB Mines Rescue Service Badge, 1957. Kerrigan passed out at Dinas Mines Rescue Station on 23 March 1957 and he was presented with the badge by Superintendent John Perry. Dinas Mines Rescue Station was opened on the 27 June 1912 through a visit of King George V and Queen Mary.

Opposite, above: New Tredegar Colliery, (The Moving Mountain Pit), Rhymney Valley, Glamorganshire in 1930. The mine was sunk in 1854. Seams worked were the Big Vein and Yard. On 12 November 1860 the accidents reports show that forty-five-year-old labourer William Shellard was injured by an explosion of firedamp and died on the 28 November. New Tredegar Colliery was abandoned in the 1920s.

Opposite, below: Elliot Colliery pit bank in the 1930s. In 1966 the Elliot manager was K. Butcher (3,207 First Class) and the undermanager was H.K. Hayes (9,154 Second Class). Seams worked were the Threequarter and Lower Four-Feet. Elliot West was closed on 1 October 1962 and Elliot East Colliery was closed on 29 April 1967 by the NCB.

Groesfaen Colliery, near Deri, Darran Valley, Glamorganshire in 1908. Groefaen Colliery was sunk in 1906 by the Rhymney Iron Co. In 1920 electric winding engines were installed. In 1925 Powell Duffryn Associated Collieries Ltd owned and worked the mine up to Vesting Day, 1 January 1947.

Groesfaen Colliery in 1913. In 1966 the Groefaen manager was W.G. Parry (5,855 First Class), assistant manager C.R. Powell (9,516 Second Class) and the undermanager was A. Evans (9,154 Second Class). Seams worked were the Seven-Feet, Red Vein, Upper Four-Feet and Lower Four-Feet. Groesfaen Colliery was closed on 22 November 1968 by the NCB.

Darran Colliery, near Deri, Darran Valley, Glamorganshire in 1909. The mine was sunk in 1869 by Rhymney Iron Co. The two shaft depth was 330ft and sunk to the Brithdir seam. On 25 February 1903 the accidents reports show that forty-five-year-old repairer Rees Jones and forty-two-year-old James Stenning were fatally injured.

Darran Colliery in 1909. The photograph shows the miners at Darren Colliery wearing the Peg and Ball naked light, prior to the explosion on 29 October 1909 which killed twenty-seven men and boys. Peg and Ball naked lights were in use at the time of the explosion. Safety lamps were used following the explosion. Darran Colliery was abandoned in 1919.

Ogilvie Colliery in 1966. In 1966 the Ogilvie manager was G. Robinson (5,811 First Class), assistant manager G. Morgan (9,070 Second Class) and the undermanagers were D.I. Roberts (8,245 First Class) and D. John (7,322 First Class). Seams worked were the Seven-Feet, Upper Four-Feet, Lower Four-Feet and Rhas Las. Ogilvie Colliery was closed on 7 March 1975 by the NCB.

Underground in 1969. At the commencement of work every miner would report to the lamp room where he would collect a lamp check bearing his number on from a board placed on the wall of the lamp room. He would then collect a lamp from the stand and place his lamp check in the vacant space. He would then report to the Lodge, the name given to the office where officials gave their instructions.

Pig Tail chewing tobacco sign. Pig Tail chewing tobacco was available at the colliery canteen for 1*s* (5p) in the 1960s. From the Lodge the miner would make his way to the top of the pit for his 'bond'. A bond is a name given to the pit cage when carrying miners. In 1871 over 34,000 miners were employed in the Glamorganshire Pits.

South Wales Colliery managers in the 1960s. From left to right: D.M.J. Evans, J.H. Jones, H. James, G. Schewitz. In 1961 the Rhymney No.5 Area General Manager was G. Tomkins, the Area Production Manager was A. Maddox and the No.1 Group Manager was J.H. Jenkins. No. 1 Group Collieries: Elliot, Ogilvie, Groesfaen; total output 814,099 tons; total manpower 3,643.

Bargoed Brithdir House Coal Pit coalface in the 1930s. Shaft depth 580ft and worked the Brithdir seam. In 1909 the Bargoed Colliery produced a world record with 4,020 tons wound in a single coaling shift. On 17 July 1913 the accidents reports show that forty-nine-year-old labourer Henry Shephard was fatally injured. Brithdir House Coal Pit was closed on 26 November 1949 by the NCB.

Opposite, above: Cilhaul Colliery, near Deri, Darran Valley, Glamorganshire in 1875. The mine was sunk in 1864. Shaft depth 90ft. Cilhaul Farm is in the background of the photograph and there are women surface workers on the right at the top of the tipping chute. In 1918 the shafts were deepened to become part of the Ogilvie Colliery. The Ogilvie Shaft was reputedly sunk around the Cilhaul Shaft. Cilhaul Colliery was closed in 1918.

Opposite, below: Bargoed Colliery, Bargoed, Rhymney Valley, Glamorganshire in 1910. The mine was opened in 1897 by Powell Duffryn Associated Collieries Ltd. Shaft depth was 1,909ft and 21ft in diameter. Seams worked were the Yard, Seven-Feet and Upper Four-Feet. The third shaft on the left in the photograph is the Bargoed Brithdir House Coal Pit. Bargoed Colliery was closed on 4 June 1977 by the NCB.

Britannia Colliery upcast shaft, Pengam, Rhymney Valley, Glamorganshire in 1984. Sinking started at Britannia Colliery in 1910 by the Powell Duffryn Associated Collieries Ltd. By the outbreak of the First World War the first coal was being raised and all plant at the mine was driven by electricity. From then, until Nationalisation it was part of the widely-known Powell Duffryn Empire and, in the peak years of coal production in the 1920s, it employed more than 1,700 miners. At the time of closure the original winding engines were still working.

In 1954 with a manpower of 1,071 it produced an annual saleable output of 330,000 tons, and in 1958 with a manpower of 1,112 it produced an annual saleable output of 398,000 tons.

In 1966 the Britannia manager was A. Craddock (3,825 First Class), assistant manager J.P.R. Evans (5,762 First Class) and the undermanagers were V. O'Neill (7,753 First Class) and W.G. Mathews (7,921 First Class). Seams worked were the Seven-Feet, Upper Four-Feet, Red Vein and Rhas Las.

Following the publication in 1973 of *Plan for Coal* a massive capital investment programme was launched by the NCB. This colliery was one of those involved in the programme.

In 1976, this fully-mechanised pit employed 800 miners and was one of a string of South Wales Collieries producing vital supplies of coking coal for the British steel industry. It worked an area of around four and half square miles. Britannia Colliery was bounded by neighbouring Oakdale and North Celynen to the east and by the twin Pengam and Penydarren Faults to the west. Across the middle of the 'take' lies the 250ft Britannia Overthrust, uniquely separating workings from the north and south shafts. A £950,000 development had involved the driving of a new 847yd main roadway through the overthrust to connect the north and south pit workings, to enable coal winding to be concentrated in the south shaft. Also a 1,000yds drift was driven to enable all men and materials to be wound in the north shaft.

Development drivages through the Pengam Fault, into a 'trough' of virgin coal lying between the Twin Faults in the Upper Nine-Feet seam, would expose more than 400,000 tons of new reserves to modern mining techniques. Coal production was in the Seven-Feet and Lower Four-Feet seams.

Britannia was one of a number of local mines which faced particularly wet mining conditions and normal daily pumping had to deal with between seven and nine million gallons of water every day.

Within the colliery area was the well-known Britannia School of Mines and the NCB's Britannia Training Centre, which had been responsible for supplying a steady stream of skilled mine workers and craftsmen to the South Wales Coalfield.

In 1976 with a manpower of 800 it produced an annual saleable output of 241,082 tons; it produced an average weekly saleable output of 3,936 tons; average output per man/shift at the coalface 5 tons 2cwt; average output per man/shift overall 1 ton 6cwt; deepest working level 1,950ft; number of coal faces that were working, 2.

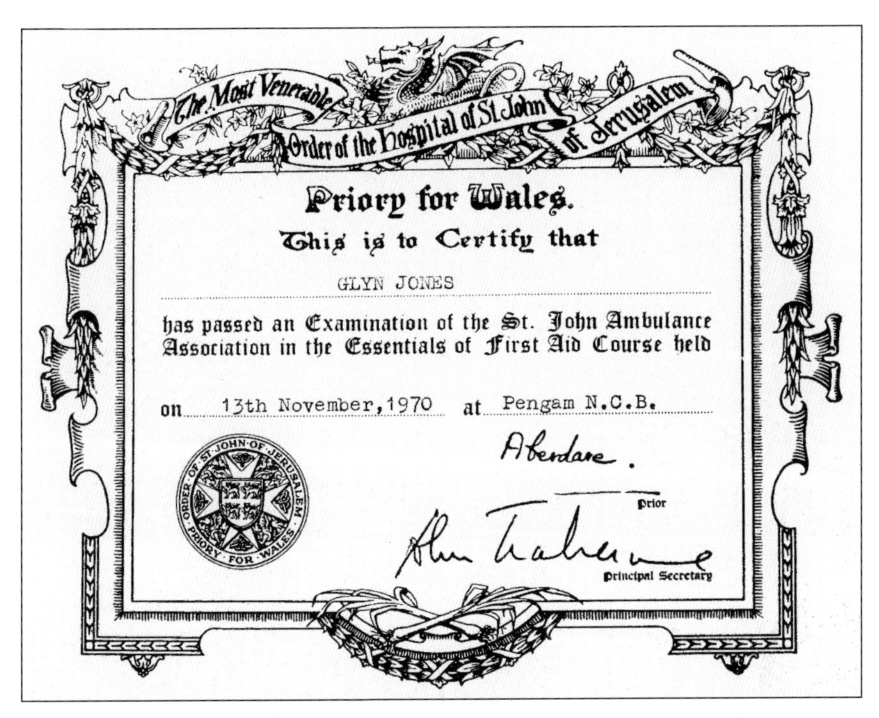

The Most Venerable Order of the Hospital of St John of Jerusalem

Priory for Wales.

This is to Certify that

GLYN JONES

has passed an Examination of the St. John Ambulance Association in the Essentials of First Aid Course held

on 13th November, 1970 at Pengam N.C.B.

Aberdare.

Prior

Principal Secretary

ORDER OF ST JOHN OF JERUSALEM PRIORY FOR WALES

Certificate awarded to Glyn Jones for passing an examination of the St John First Aid course at Britannia School of Mines on 13 November 1970. In 1976 the Britannia No.1 shaft depth was 1,844ft, diameter 21ft; No.2 shaft depth 2,142ft, diameter 21ft; manwinding capacity per cage wind 25/50; coal winding capacity per cage wind 6 tons; winding engines horsepower 1,350/2,125; stocking capacity on pit surface 65,000 tons; average weekly washery throughput 14,000 tons; types of coal prime coking blend; markets, steel; fan capacity, cubic feet per minute 250,000; average maximum demand of electrical power 6,200kW; total capital value of plant and machinery in use was £543,000; estimated workable coal reserves 4.8 million tons.

The colliery was once described as a pit where nothing ever happened apart from raising coal and pumping minewater every day.

On 22 August 1914 the accidents records (1910/1914) show that forty-year-old timberman Patrick Witty was fatally injured.

By the late 1970s the workforce of 800 men raised an annual saleable output of 241,108 tons of coal from two faces in a take of four square miles.

Britannia Colliery Coal Seams:

Depth Feet	*Standard Name*	*Local Name*
883	Brithdir	Brithdir
1,845	Gorllwyn	Gorllwyn
1,940	Two-Feet-Nine	Elled
1,998	Four-Feet	Upper Four-Feet
2,022	Six-Feet	Big Vein
2,130	Upper Nine-Feet	Red Vein
2,138	Lower Nine-Feet	Rhas Las
2,240	Amman/Yard/Seven-Feet	Amman/Seven-Feet Rider/Seven Feet
2,330	Five-Feet/Gellideg	Lower Four-Feet

Britannia Colliery was closed on 8 December1983 by the NCB.

A Paragon large sterilized plain wound dressing used in 1930. In 1954 First Aid boxes were carried by qualified First Aid men on each shift, one box for every thirty persons. The contents of the box were: three large mine dressings, six small mine dressings, six finger dressings, some burn dressings, a supply of iodine ampoules, one tourniquet and some triangular bandages, a pencil and notebook to enter casualty details etc.

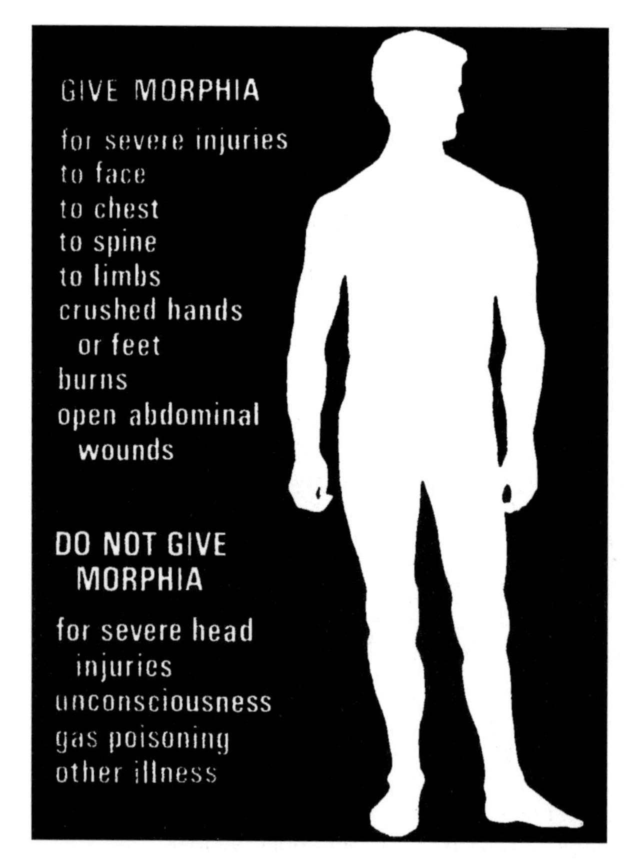

Left: Morphia instructions: 'Give morphia for severe injuries to face; to chest; to spine; to limbs; to crushed hands or feet; burns; open abdominal wounds. Do not give morphia for severe head injuries; unconsciousness; gas poisoning; other illness.'

Right: Emergency action in cases of electric shock. 'Contact with electric current: 1. Damage to heart muscle, which might stop it beating and/or 2. Temporary paralyses of chest muscles causing breathing to stop and/or 3. Burns. Priority one: Break the contact between the casualty and the current; by switching off power, or removing plug, or wrenching cable free. If this is impossible: by means of an insulating material, move the casualty away from contact. Always be certain to stand on dry insulating material.'

Emergency Action in Cases of Electric Shock

The notes on this card are intended to remind you of the steps to be taken in the event of electric shock. They are not a substitute for a course of training in first aid details of which can be obtained from your training officer.

CONTACT WITH ELECTRIC CURRENT

1. Damage to heart muscle, which might stop it beating and/or
2. Temporary paralysis of chest muscles causing breathing to stop and/or
3. Burns.

Priority One

Break the contact between the casualty and the current:

by switching off power, or
removing plug, or
wrenching cable free.

If this is impossible:

by means of an insulating material, move the casualty away from contact.

Always be certain to stand on dry insulating material

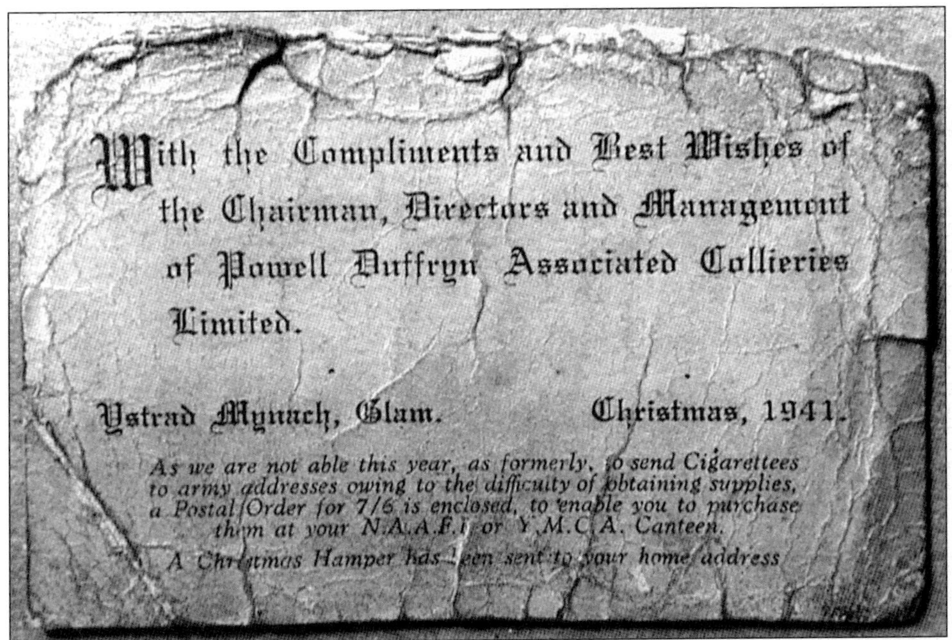

With the Compliments and Best Wishes of the Chairman, Directors and Management of Powell Duffryn Associated Collieries Limited.

Ystrad Mynach, Glam. Christmas, 1941.

As we are not able this year, as formerly, to send Cigarettees to army addresses owing to the difficulty of obtaining supplies, a Postal Order for 7/6 is enclosed, to enable you to purchase them at your N.A.A.F.I. or Y.M.C.A. Canteen.

A Christmas Hamper has been sent to your home address

A message from Powell Duffryn Associated Collieries Ltd in 1941. 'With the Compliments and Best Wishes of the Chairman, Directors and Management of Powell Duffryn Associated Collieries Limited, Ystrad Mynach, Glam. As we are not able this year, as formerly, to send Cigarettes to army addresses owing to the difficulty of obtaining supplies, a Postal Order for 7/6 (37½p) is enclosed, to enable you to purchase them at your NAAFI or YMCA Canteens. A Christmas Hamper has been sent to your home address.'

'The wife's away for the day'

Opposite, above: A colliery newspaper cartoon by Richard Hughes in 1947. 'The wife's away for the day'. The NCB *Coal* magazines and the *Coal News* were popular with the miners and often available and read at the colliery canteen. In 1947 the *Coal* newspaper was 4*d*.

Opposite, below: Penallta No.1 and No.2 Colliery, Hengoed, Rhymney Valley, Glamorganshire in 1991. Sunk by the Powell Duffryn Empire, during their rapid expansion in the early 1900s, the Penallta shafts were put down in 1906 and 1907 and were completed by 1909. The location of the mine allowed adequate space for the company to lay out one of the most modern colliery surfaces of the day and Penallta quickly became one of Powell Duffryn's most productive and profitable pits.

A typical Powell Duffryn pit which, although impressive with its vast central engine house for winding engines, compressors and generators, was not particularly attractive when compared to Deep Navigation Colliery or Deep Duffryn Colliery.

By the early 1930s, the pit employed over 3,000 miners and was showing an annual production figure of 860,000 tons, all top quality steam coal. By 1935, Penallta was the European record holder for windings of coal in a week.

Immediately after Nationalisation in 1947, the pit also led the way, by becoming one of the first in the South Wales Coalfield to install a Meco-Moore Cutter-Loader, one of the first power loading machines to be adopted in British collieries.

Electrification in the early 1960s replaced the older steam winding gear and formed the vanguard of a £2½ million modernisation project, which also deepened the mine to new working levels.

In 1966 the Penallta manager was W.J. Llewellyn (8,488 First Class), assistant manager K. Morgan (9,124 First Class) and the undermanager was M. Brocket (6,974 First Class). Seams worked were the Red Vein, Rhas Las and Seven-Feet.

Following the publication in 1973 of *Plan for Coal* a massive capital investment programme was launched by the NCB. This colliery was one of those involved in the programme.

In 1976 the mining programme at Penallta worked an area of around four square miles, bounded on the west by the Gelligaer Fault and, on the east by the equally impressive Penydarren Fault, both causing a massive 300ft 'throw' in the continuation of the coal seams. To the north and south lie the abandoned workings of the closed Groesfaen Colliery (closed 22 November 1968) and Llanbradach Colliery (closed 29 December 1961).

Penallta Emperor a sixteen- hand dark-brown gelding, pictured here in 1938. In 1976 all coaling was in the Seven-Feet seam, at a depth of 2,400ft and the operation involved almost eight miles of underground roadways and more than two miles of high-speed belt conveyors. The workings were amongst the deepest in South Wales and the coalfaces were an average mile and a half in from the shaft bottoms. Over half of this was covered by loco-hauled manriding trains which, during shift times, doubled as material transport. With a manpower of 690 Penallta Colliery produced an annual saleable output of 218,682 tons; it produced an average weekly saleable output of 4,921 tons; average output per man/shift at the coalface 5 tons 8cwt; average output per man/shift overall 1 ton 14cwt; deepest working level 2,400ft; number of coal faces that were working, 2; No.1 shaft depth 2,400ft, diameter 21ft; No.2 shaft depth 1,955ft, diameter 21ft; manwinding capacity per cage wind, 50; coal winding capacity per cage wind 6 tons; winding engines horsepower 1,890; stocking capacity on pit surface 500,000 tons; average weekly washery throughput 8,000 tons; types of coal blended; markets, power station; fan capacity, cubic feet per minute 266/295,000; average maximum demand of electrical power 4,400kW; total capital value of plant and machinery in use was £530,456; estimated workable coal reserves 3.8 million tons.

£3½ million was allocated to Penallta Colliery at the end of 1985 for a heavy-duty high-technology coalface. In the same year plans were made to install rapid coal winding skips (a container to carry coal up the shaft) which it was hoped would raise shaft capacity by thirty-three per cent.

Despite of the new equipment and investment the colliery ceased working in 1991.

Penallta Colliery Coal Seams:

Depth Feet	*Standard Name*	*Local Name*
2,025	Two-Feet-Nine	Two-Feet-Nine
2,045	Four-Feet	Four-Feet
2,115	Six-Feet	Six-Feet
2,195	Upper Nine-Feet	Red
2,210	Lower Nine-Feet	Rhas Las
2,305	Yard & Seven-Feet	Seven-Feet

Penallta Colliery methanometer. A methanometer is a modern device for the detection of methane gas in a mine, and is carried by a fireman who is a management official in the coal mines and is employed to carry out safety inspections, especially to detect the presence of gas. An explosion is caused by the ignition of combustible firedamp gas, (a mixture of methane (predominant), carbon monoxide, nitrogen, ethane, carbon dioxide) and air.

House coal hauliers Royston Jones and Phillip Davies, 9 July 1999. Roy and Phil deliver concessionary coal to ex-miners, widows etc. to their homes in the valley. Firedamp and air mixtures can only be exploded within a definite range. The lower limit of the range is 5.4 per cent and ends at the upper limit of 14.8 per cent. There is a greater chance of an explosion when the percentage is about 9 per cent methane and air. Penallta Colliery was closed in October 1991 by the NCB.

Universal Colliery miners in the 1930s. Universal Colliery Senghenydd, Aber Valley, Glamorganshire was sunk by the Universal Steam Coal Co. in 1891. The two shafts' depth was 1,950ft. Seams worked were the Two-Feet-Nine, Four-Feet, Six-Feet and Nine-Feet.

In 1905 the Lewis Merthyr Consolidated Collieries Ltd acquired the Universal Colliery, which was later to suffer the worst-ever mining disaster in British history. The first of two dreadful explosions struck at 5:00 a.m. on 24 May 1901 when 82 men and boys were killed in the east side of the mine about 700yds from pit bottom. Although the ventilation continued to be very good, the heavy and extensive falls of roof greatly hampered rescue efforts. At pit bottom haulier William Harris was found unconscious beside the body of his dead horse. He was the only one to be brought up alive.

The second explosion occurred at Universal Colliery on Tuesday 14 October 1913, the worst mine disaster in the history of the British Coalfield when 439 men and boys were killed (437 killed by the explosion and 2 rescuers killed by falls). A sad testament to tragedy on a scale unimagined in the annals of coalmining in Britain can be found etched in the tombstones in the Aber Valley village graveyards of Eglwysilan and Penyrheol. There, the chilling epitaphs bear witness to a legacy of grief that devastated a small valley community and record a dark chapter in the history of coal mining in Glamorganshire. The inscriptions read: 'Bu farw yn Nanchwa Senghenydd'. 'Died in the Senghenydd Explosion'.

On the morning of Tuesday 14 October 1913, the day shift assembled to descend the downcast shaft of Lancaster Pit. Because of the size of the workforce, which was around 950 men and boys, it was 8:00 a.m. before work in most of the districts was in progress. Within ten minutes, disaster was to strike on a scale previously unimagined. A terrific explosion occurred, followed by a blast which roared up the Lancaster pit, sweeping the cage before it. Within seconds, the carriage was hurled into the winding gear, smashing the wooden platform into a mass of splintered wreckage.

The blast shook the village to its foundations, and could be heard as far away as Risca. No one there could believe it had happened.

Colliery manager Edward Shaw rushed to investigate and was shocked by the devastation. His immediate concern was the ventilation fan. Relieved to find it undamaged he turned his attention to the removal and replacement of the broken cage at the top of the downcast shaft.

Soon, villagers came rushing to the pit, summoned by the sound of the blast. After assembling a party of volunteers, Shaw together with two officials and several nightshift miners began their descent. As the rescuers left the cage, voices could be heard from the Nine-Feet seam a further 60ft below. Moving forward the rescuers made their way down the return road in the hope that they could use the west cross-cut to gain access into the Lancaster workings. All too soon their way was barred by dense choking smoke and an alternative means of entry through the east cross-cut was sought. With grim determination they battled through the heat and smoke and at last managed to gain access to the Lancaster shaft. Beyond the point where the arches ceased it was all double timbering and every pair of timbers so far as could be seen was on fire up to the entrance of No.1 North. It was like looking into a furnace.

Shaw knew that it would take several hours to reverse the fan to enable smoke and gases to be drawn out of the workings. This inability to quickly reverse the ventilating system was to provide damning evidence against the colliery manager and owners. Shaw returned to the surface to seek outside help. The rescue teams from Dinas, Aberdare, the Rhymney Valley and Crumlin were soon on the scene.

By Monday 20 October fires in No.1 North were under control, but progress in the main west level continued to be hampered by fire, fumes and falls. Rescuers continued to work undaunted by the fact that more than one and a half miles still remained between them and the area where their trapped comrades lay.

A little over a month after the disaster, a statement was issued giving the final death toll as 439, of which 406 bodies had been recovered and thirty-three still remained unlocated within the mine. Of the number brought out, only 346 could be positively identified.

Acting with some expediency the Inspector of Mines wrote to all collieries in South Wales: 'The explosion was proved beyond all reasonable doubt to have been caused by the sparking of an electrical signalling bell which ignited an accumulation of gas ... I am instructed therefore to call your attention to the necessity of strict observance of electrical special rule 15 (1) with reference to signalling apparatus'.

On Tuesday 14 October 1913 at approximately 8.10 a.m. one of the saddest chapters in the history of the South Wales Coalfield came to a close. It was a chapter marked with the tragic loss of 439 men and boys and the immeasurable grief of an entire valley community. Universal Colliery ceased production in 1928 and the shafts were filled and capped with a concrete seal in 1979.

Opposite, above: A section of coalface at Windsor Colliery in 1966. Windsor Colliery later became part of the famous Powell Duffryn Empire, where it remained until Nationalisation. In 1954 with a manpower of 1,073 it produced an annual output of 299,000 tons and in 1956 with a manpower of 1,067 it produced an annual output of 288,000 tons. In 1966 the Windsor manager was W.J. Watkins (8,089 First Class) and the undermanagers were K.J. Davies (7,494 First Class) and B. Preece (7,937 First Class). In 1976 the combined Windsor/Nantgarw mine and its washery employed around 654 men and produced over 4,000 tons of high quality coking coal a week. (The two collieries were merged in 1975). Seams worked were the Six-Feet (Four-Feet) and Amman Rider/Yard/Seven-Feet (Nine-Feet). Windsor/Nantgarw Colliery was closed on 7 November 1986 by British Coal.

Windsor Colliery, Abertridwr, Aber Valley, Glamorganshire in 1913. The first coal came up the shaft of Windsor Colliery in 1902, as Britain celebrated Queen Victoria's Diamond Jubilee. Sinking started five years earlier by the Windsor Steam Coal Co. and, completed to a depth of 2,018ft, made the pit one of the deepest in the South Wales Coalfield at the time.

Right: Black Brook Colliery, near Caerphilly, Glamorganshire in 1928. The mine was sunk in 1921 and owned by J. Morgan, a publican and brewer. Seams worked were the Bute, Rhondda No.1 and Rhondda No.2. Black Brook Colliery was closed on 7 November 1986 by British Coal. In 1871 the Amalgamated Association of Miners formed a strong union among the miners of South Wales. In 1872 new legislation was introduced to regulate the operation of 420 coal mines in the South Wales Coalfield. In 1873 the South Wales Coal Owners Association was formed. In 1875 the sliding wage scale was introduced to determine the level of wages among the miners. In 1898 miners were locked out and the South Wales Coalfield was at a standstill; after six months the miners were defeated. The South Wales Miners' Federation was formed.

Rhos Llantwit Colliery, near Caerphilly, Glamorganshire in 1895. A flat winding rope can be clearly seen in the background. There is a canopy over the cage indicating a wet shaft. On 23 August 1888 the accidents reports show that thirty-seven-year-old collier James Drew was killed by a fall of stone. Rhos Llantwit Colliery was abandoned in September 1902 by the Rhos Llantwit Coal Co.

Opposite, above: Bedwas Navigation Colliery, Bedwas, near Caerphilly, Glamorganshire in 1960. The first coal reached the market in 1912. In 1923 with manpower of 2,300 it produced an annual saleable output of 500,000 tons. In 1976 the deepest working level was 2,400ft and the number of coal faces working was three. Bedwas Navigation Colliery was closed on 31 August 1985 by British Coal.

Old Pentwyn Pit, Machen, near Caerphilly, Glamorganshire in 1920. The person standing extreme right in the photograph with a stick and leather hat was the manager or undermanager and the lady in the back row was probably the office cleaner. Old Pentwyn Pit was abandoned in 1922.

In Everlasting Memory to the Miners who lost their lives from Rhymney to Caerphilly

Date	Mine	Lives Lost
18 October 1852	New Place, Rhymney	Hitcher Daniel Williams (31) and collier David Jones (22) killed by falling down the shaft. Tooth of pinion of winding engine broke.
15 June 1853	Number One Coal Level	Collier John Evans (14) killed by explosion of firedamp.
7 September 1853	Duffryn, Rhymney	Doorboy David Griffiths (11) killed when he was crushed between drams.
21 January 1860	Coedcaeddu, Rhymney	Collier John Morgan (26) killed helping a fellow workman to pinch down some shale; he unfortunately knocked out a prop, which was immediately followed by a heavy fall of coal. There was a complete triangle of slips through which the mass fell.
21 February 1868	Tyr Phil, Rhymney	Collier John Jones (22) hurt in the head by a fall of shale and was so badly injured that he died on the 22nd.
26 February 1868	Mardy, Rhymney	Collier John Jones (18) killed when he fell whilst drawing away some coal at the bottom of a 'slip'.
23 March 1868	Terrace, Rhymney	Collier Daniel McArthy (40) killed whilst he was ramming the safety fuse and the shot went off. It is supposed that he had in a blundering manner placed his light close to the shot hole and so fired the fuse.
22 June 1876	Bargoed, Rhymney	Collier Taliesin Rees (13) killed by a fall of stone.
7 June 1884	Elliot, Rhymney	Sinker Gwyliam Williams (21) killed.
13 March 188	Llanbradach Pwllypant	Sinker William Green killed when an iron bar fastened to the headstock for the automatic emptying of the water barrel was, in consequence of overwinding, pulled off and then fell down the shaft and struck him.
19 May 1888	Rudry, near Caerphilly	Collier Thomas Righton (21) killed by a piece of coal which he was pulling from the face pressed the handle of his pick against his side. He filled his tub of coal, walked home a distance of two or three miles and died the same evening.
27 November 1890	Bryngwyn, Bedwas	Collier Eli Tanner (48) killed.
3 September 1902	McLaren, Rhymney	Sixteen miners killed and twenty-one injured by an explosion of firedamp in the Yard Vein.
12 May 1903	Elliot, Rhymney	Collier's boy David John Vowles (18) killed.
18 December 1914	Bedwas, Caerphilly	Collier Thomas Evans (42) collier's helper William Griffiths (41) killed.

Once again a sad reminder of the true price of coal.

Three

Tredegar to Cwmfelinfach in the South Wales Coalfield

With such rapid growth in the coal industry it was inevitable that great wealth was accumulated by the coal owners. These men, among the richest in the country, were with a few exceptions, hard task masters. They often looked upon their workers as objects rather than human beings, and it was generally their belief that the men were responsible for explosions underground which occurred with terrifying regularity throughout the South Wales Coalfield. The owners believed that smoking underground was the major cause of accidents and prosecuted any miner found infringing regulations, but they were themselves powerful enough to evade those few cases laid against them for compensating accident victims.

From the outset, a sharp division existed between 'master' and 'workmen'. While the former lived in splendid luxury, the worker existed on the margins of poverty. The owners openly expressed their belief that if kept on low wages, the miner would work harder to supply his family's needs. If paid high wages, they claimed, the men became lax in their habits and squandered their money on drink.

When the workers began joining the newly-formed trade unions their employers countered with threats of the sack. To endorse this they began issuing unfavourable discharge notes to those who spoke out against poor working conditions. To obtain work at any pit in the South Wales Coalfield, a collier had to produce a discharge note from his previous employer stating that he was an industrious and trustworthy person. It followed that any miner sacked and given no such note would be unable to find other work.

In 1873, the Coal Owners Association was formed by leading magnates of the industry. This organisation provided a powerful voice in dealing with demands for better working conditions or higher rates of pay. The association was to become notorious for its sliding wage-scale which protected coal owners from any loss of profits from depressed world prices. To ensure the profit margins were kept high, miners wages were lowered whenever a fall in the demand for coal occurred. Inevitably, such a system led to a great deal of bitterness among the workers and it was not surprising therefore that the miners opposed the introduction of any new work patterns, such as a double shift system which was common practice in many other areas.

By the turn of the century new forces began to appear in the coalfield. Many of the coal owners united into larger companies. The mighty Powell Duffryn Empire had, by various mergers and purchases, come to control over seventy-five pits. In 1914 this group issued figures, which put the reserves in the Rhymney and Aberdare Valley as in excess of 448,000,000 tons. Such was the confidence in the industry at the time that brave forecasts were made that mining operations would continue on the same massive scale for the next hundred years. Already though, the clouds of war were gathering and although bringing massive profits to owners, it was to usher a new age of conflict between employers and miners.

Strikers' Levels, Tredegar, Sirhowy Valley, Monmouthshire in 1926. The striking miners are seen at the entrance of the levels they had opened to obtain coal for their own domestic use during the 1926 strike. The Strikers' Levels had a chequered history of openings and closures.

Ty Trist Colliery, Tredegar, Sirhowy Valley, Monmouthshire in 1956. No.1 and No.2 shafts were sunk in 1834 and No.3 shaft was sunk in 1868. Seams worked were the Big Vein, Yard, Old Coal and Meadow Vein. In 1954 with a manpower of 354 it produced an annual output of 66,000 tons.

Opposite, below: Whitworth Colliery, Tredegar, Sirhowy Valley, Monmouthshire in 1900. The mine was opened in 1876 by the Tredegar Iron & Coal Co. Seams worked were the Old Coal and Bydelog. The Pit merged with Ty Trist Colliery. Whitworth Colliery was closed with Ty Trist Colliery on 31 January 1959 by the NCB.

Dram Tipplers (tumblers) in 1954. On top of the pit at the Pit Bank the tipplers would invert and tip the drams of coal onto a moving conveyor directly underneath and convey the minerals into the washery for cleaning, washing and grading the coal. Ty Trist Colliery was closed on 31 January 1959 by the NCB.

Opposite, above: This poor quality reproduction shows Pochin Colliery, near Tredegar, Sirhowy Valley, Monmouthshire in 1950. The mine was sunk in 1876 by the Tredegar Iron Co. It was abandoned for four years and finally completed in 1880. Seams worked were the Big Coal, Yard, Meadow Vein and Upper Rhas Las.

Left: Miners in the bond ready to descend the shaft in 1910.

Pochin Colliery miners proudly present their trophies of black gold (Aur Du) in 1913. In 1954 with a manpower of 693 it produced an annual output of 152,000 tons and in 1958 with a manpower of 646 it produced an annual output of 116,000 tons. Pochin Colliery was closed on 25 July 1964 by the NCB.

Opposite, below: Markham Colliery miners in 1920. Both shafts – the north upcast and south downcast shaft – were 18ft in diameter and were sunk to the Old Coal seam at a depth of 1,800ft the headframes were of lattice steel and 65ft high; the cages were originally single deck with two drams on the deck.

Right: Markham Colliery, Markham, Sirhowy Valley, Monmouthshire in 1984. Markham Colliery was situated in the Sirhowy Valley about four miles to the south of Tredegar. The sinking of the two shafts commenced in 1910 by the Markham Steam Coal Co., a subsidiary of the Tredegar Steel, Iron and Coal Co. The colliery was named after Sir Arthur Markham the company chairman. On 18 May 1912 the accidents reports show that mechanic George Jones (25), fitter Michael Carroll (33), fitter Albert Leggett (24), banksman Thomas Patrick (44) and sinker George Guntrip (27) were killed by an explosion.

By 1913 the first supplies of high-grade coking coal were leaving the pit. During the sinking, ten clearly defined coal seams were encountered and all but two of these have been worked during the life of the mine.

Markham M18 coalface cutterman Gwyn Roderick in 1978. The colliery was entirely electrically powered, one of the earliest pits to be worked entirely by this means. The motors on the winding engines were made by Siemens Schuckert and the mechanical portion was by Markham & Co. of Chesterfield. The winding engine drums were semi-conical and 12ft to 18ft in diameter. The colliery was ventilated by two Waddle fans of 18ft diameter each producing 400,000 cubic feet of air per minute. The Upper Rhas Las seam was also worked by a conveyor system and had been worked thus since 1920. Pneumatic picks were used on the seam. Trunk conveyors were installed in April 1946 and all coal was worked on the long-wall system. Seams worked in 1946 were the Old Coal of which fifty per cent was machine cut. The Big Vein was worked extensively throughout the take. The Yard was worked by machine cutting and conveying.

At Nationalisation, the workforce was 1,107 men underground, with 196 men working on the surface. By the 1950s coal was wound in both shafts which were capable of raising 15,000 tons per week. In 1954 with a manpower of 1,356 it produced an annual output of 263,000 tons; in 1956 with a manpower of 1,259 it produced an annual output of 301,000 tons and in 1958 with a manpower of 1,189 it produced an annual output of 276,000 tons.

Following the publication in 1973 of *Plan for Coal* a massive capital investment programme was launched by the NCB. This colliery was one of those involved in the programme. In 1976 production was concentrated in the Yard/Meadow Vein seam at a depth of 2,088ft and the Five-Feet/Gellideg seam at 2,140. The collieries 'take' was an area of around 1.8 square miles, bounded on the east by the 135ft Transport Fault and on the other three sides by workings of other collieries; Pochin (closed 25 July 1964) to the north; Elliot (Elliot West was closed on 1 October 1962 and Elliot East Colliery was closed on 29 April 1967) to the west and Oakdale (closed August 1989) to the south. The mining programme at Markham Colliery involved around ten miles of underground roadways, carrying almost 3¼ miles of high-speed belt conveyors. However, a major development that was under way would add a new length of roadway which, although only four tenths of a mile long, would prove to be one of the pits most vital elements. The development costing £9 million would link Markham and neighbouring North Celynen to Oakdale and its massive coal preparation plant. Markham coals will then be taken underground, direct to Oakdale Colliery for washing and grading, eliminating a surface rail link of over two miles.

From left to right: Mr John Perret, Manager; Mr Davies, Director of Machinery; Gwyn Roderick (cutterman); Brian Evans; Trevor Williams (asst cutterman); Tom Morgan (craftsman), in 1978. The photograph was taken when the first electric remote control cutter in the South Wales Coalfield was installed at the pit in the M18 Meadow Vein.

Meanwhile, the 646 miners at Markham continue to make a valuable contribution to the annual flow of essential coking coals from the South Wales Coalfield, the country's finest, with an average annul output in excess of 230,000 tons. With a manpower of 646 it produced an annual saleable output of 225,325 tons; it produced an average weekly saleable output of 3,936 tons; average output per man/shift at the coalface 6 tons; average output per man/shift overall 1 ton; deepest working level 1,800ft; number of coal faces that were working, 3; No.1 shaft depth 1,845ft, diameter 18ft; No.2 shaft depth 1,854ft, diameter 18ft; manwinding capacity per cage wind, 24; coal winding capacity per cage wind, 3 tons; winding engines horsepower 1,140; types of coal, coking; markets, steel; fan capacity, cubic feet per minute 400,000; average maximum demand of electrical power 3,965kW; total capital value of plant and machinery in use was £331,117; estimated workable coal reserves 2.6 million tons.

The first coalface lighting to be installed on a face in the South Wales Coalfield in 1977 with seventy face lighting units were installed on the M22 face, but damage occurred from falling stones and forty units had to be returned to the manufacture for repair. When working, the system was very impressive and was a credit to the face team.

By the end of the 1970s the colliery had around ten miles of underground roadway with just over three miles of high-speed belt conveyors. By 1980 a major development linked Markham and the adjacent North Celynen (closed 31 March 1985) to Oakdale and Markham coal was taken directly underground to Oakdale for washing and grading. This three-pit complex was then the biggest single production unit in the South Wales Coalfield.

In December 1982 the men at Markham produced 8,504 tons of saleable coal in one week after a period of working in poor geological conditions. The 682ft-long face was one and a half miles from pit bottom and advanced 38ft in one week. At this period the colliery employed 615 men working on two faces.

Markham Colliery Coal Seams

Depth Feet	*Standard Name*	*Local Name*
1,393	Two-Feet-Nine	Elled
1,439	Four-Feet	Big Vein
1,446	Upper Six-Feet	Yard Vein
1,500	Lower Six-Feet	Threequarter
1,539	Upper Nine-Feet	Polka
1,571	Lower Nine-Feet	Rhas Las
1,725	No.2 Yard & Seven-Feet	Meadow Vein
1,799	Five-Feet & Gellideg	Old Coal

Markham Colliery was closed on 20 September 1985 by the NCB.

Oakdale Colliery, Oakdale, Sirhowy Valley, Monmouthshire in 1908. Sinking started at Oakdale in 1908 and, by 1910 the first supplies of high grade coking coal were being raised by the then owners, Oakdale Navigation Collieries. Within a few years, the pit established a reputation of being one of the most productive and profitable in the South Wales Coalfield. Working an area of around four square miles, operations were bounded on the north by the 150ft Transport Fault, on the south by the 250ft Britannia Overthrust Fault and by the Britannia Colliery workings to the west. Coal was being mined from the Meadow Vein (Yard/Seven-Feet) the Old Coal (Five-Feet/Gellideg) seams at a maximum depth of 2,180ft. The working area was overlain by the old workings of the Tillery seams, draining a mighty 4.3 million gallons of water each day which had to be pumped to the surface via two main pumping stations. A proportion of the water was used by the Coal Preparation Plant and one million gallons per day was taken by the Welsh Water Authority for public use.

The headframe of the South Pit was built by Rees & Kirby at a height of 76ft to the centre pulleys which were 20ft in diameter. The steam winding engine on the South Pit was built by Markham & Co. of Chesterfield with 36in cylinders and a 7ft stroke. A semi-conical drum with a diameter of 27ft was installed in 1945. The cages on the South Pit were, in 1945, double deck each carrying two drams, each dram carrying 25cwt of coal. The average weight raised per wind was about 5 tons.

The North Pit headframe was originally 83ft high with a steam winding engine by Messrs Markham & Co. It too had 36in-diameter cylinders with a 7ft stroke, although the drum was 28ft in diameter.

The head frame on the Waterloo shaft was only 55ft high with single-deck cages for two drams each. The engine house on the North Pit was built of ferro-concrete, and the pit boxing was similarly constructed. The colliery had twelve Babcock & Wilcox double-drum water tube boilers, with a working pressure of 200psi. A power-house with a large water cooling tower dominated the colliery surface. Two Scirocco fans each 154in in diameter produced 600,000 cubic feet of air per minute.

In 1959 for the third time, water burst into the workings in the Old Coal and the colliery flooded, the electricity supply to the pumps was lost and it was only the diversion of the water into a heading, while the pumps were not working, which saved the colliery.

Oakdale colliers in the Bond in 1920. In 1967 the Oakdale manager was R.J. Williamson and the undermanager was W.J.C. Pritchard. Seams worked were the Big Vein, Meadow Vein and the Old Coal.

Following the publication in 1973 of 'Plan for Coal' a massive capital investment programme was launched by the NCB. This colliery was one of those involved in the programme.

In 1976 with a manpower of 1,020 it produced an annual saleable output of 383,838 tons; it produced an average weekly saleable output of 8,365 tons; average output per man/shift at the coalface 9 tons 3cwt; average output per man/shift overall 2 tons 1cwt; deepest working level 2,181ft; number of coal faces that were working, 3; No.1 shaft depth 2,009ft, diameter 21ft; No.2 shaft depth 2,084ft, diameter 21ft; manwinding capacity per cage wind 25/50; coal winding capacity per skip wind 10½tons; winding engines horsepower 2,250/2,500; weekly washery throughput 40,000 tons; types of coal coking; markets industrial cokes/power station; fan capacity, cubic feet per minute 380,000; average maximum demand of electrical power 5,489kW; total capital value of plant and machinery in use was £3.3 million.

A major reorganisation project was completed in 1980 at a cost of £12.5 million, linking neighbouring Markham and Celynen North Collieries to Oakdale Colliery by underground roadways. Fourteen miles of high-speed, computer-controlled conveyors carried coal from seven faces being worked in the three collieries, to the Oakdale shaft. 10½-ton skips wound the coal, at the rate of 430 tons per hour, to the Coal Preparation Plant. After preparation, the large coal was conveyed to a landsales terminal for domestic use, the remainder to a 750-ton bunker for rapid loading into British Rail wagons for onward shipment.

To extend and secure the life of the new triple unit, a further new development was authorised in 1979. This was for the driving of access roadways into an estimated 9.8 million tons of coal reserves to the south of Oakdale Colliery, near the closed Wyllie Colliery. The scheme included 1.8 miles of locomotive haulage facilities and 1.1 miles of trunk, conveyor roadways, to establish the first production face in the new area. Final completion of the project was scheduled for 1983.

Ian Roderick Deputy Stores Manager in 1989.

Oakdale Colliery:

Depth Feet	*Standard Name*	*Local Name*
1,920ft	Upper Four-Feet	Big Vein
1,968ft	Upper Six-Feet	Yard Vein
2,025ft	Nine-Feet	Rhas Las
2,136ft	Yard & Seven-Feet	Meadow Vein
2,180ft	Five-Feet & Gellideg	Old Coal

The coal mine is were the miners work, digging, pulling, pushing and loading, going down they look like light, coming up they look like night, black, sweaty, muddy and dusty.
The Coal Mine is where the miners work, carrying tools, lamps and helmets. Every time when they go home, wives they shout, argue and moan, *about the black foot prints in their home.*
The Coal Mine is where the miners work, underground their guided by their light, under great big valley mountains. My grandad told me all these things; I bet it was a hard life, digging, pulling, pushing and loading, that was the miners' life.
Melanie Roderick age twelve.

Oakdale Colliery Stores in 1989. *From left to right*: David Newton, Alwyn Davies, Martin Perret, Colin Davies. With reserves estimated at more than 22 million tons the new complex was numbered amongst the coalfield's long-life producers. With a combined annual output target of 900,000 tons, it was also the largest in the South Wales Coalfield. Oakdale Colliery was closed in August 1989 by British Coal.

Left: One of the miners' favourite safety helmets; the working conditions of water and heat would quickly cause the helmet to fit snugly to the shape of the miner's head.

Primitive coal processing: Hand washing in the washery (a modern coal preparation plant usually found at most pits). Sometimes called the Boris Kharloff dungeon because of the diabolical working conditions.

Rock Colliery, Blackwood, Sirhowy Valley, Monmouthshire in 1900. The mine was sunk in 1880 by Budds of Newport. In 1954 with a manpower of seventy it produced an annual saleable output of 17,000 tons and in 1956 with a manpower of sixty-nine it produced an annual saleable output of 16,000 tons. Rock Colliery was closed on 14 June 1957 by the NCB.

Opposite, above: Wyllie Colliery, Pontllanfraith, Sirhowy Valley, Monmouthshire in 1967. The mine was sunk in 1925 by Tredegar (Southern) Collieries Ltd, a subsidiary of the Tredegar Iron and Coal Co. Ltd. In 1954 with a manpower of 863 it produced an annual saleable output of 213,000 tons. Wyllie Colliery was closed on 21 March 1968 by the NCB.

Opposite, below: Nine Mile Point Colliery also known as Coronation Colliery, near Cwmfelinfach, Sirhowy Valley, Monmouthshire in 1968. The mine was opened in 1905 by Burnyeat & Brown. Seams worked were the Bute, Four-Feet, Yard/Seven-Feet and Black Vein. Coronation Colliery was closed on 25 July 1964 by the NCB.

OCEAN
OCEAN

In Everlasting Memory to the Miners who lost their lives from Tredegar to Cwmfelinfach.

Date	Mine	Lives Lost
4 July 1852	Cwmrhos, Tredegar	Collier John Jones (49) killed by fall of clod.
22 July 1852	Ashtree, Tredegar	Collier Thomas Brown (40) killed by clod falling whilst ripping back.
11 December 1852	New Pit, Tredegar	Collier Daniel James (55) killed by fall of roof from slip.
12 March 1853	Bedwelty, Tredegar	Collier William Jones (14) killed by fall of coal from a slip.
27 March 1853	Globe, Tredegar	Collier George Williams (26) killed by falling under drams.
9 May 1853	Ty Trist, Tredegar	Collier Rees Price (40) killed by blasting pricker struck the rock.
24 June 1853	Yard Level, Tredegar	Collier Thomas Halton (26) killed by fall of coal in the stall.
19 July 1853	Shop Level, Tredegar	Doorboy Thomas Griffiths (10) killed by fall of roof.
2 March 1860	No.8 Pit, Tredegar	Collier Henry George (29) killed when drawing back his stall in the Threequarter seam. He knocked out a prop and the upper coal fell on him.
8 April 1874	Tredegar	Haulier John Davies (12) killed when a horse ran away and a dram was upset.
30 June 1874	Forge, Tredegar	Doorboy Edward Griffith (12) killed by fall of coal.
17 August 1874	Ashtree, Tredegar	Collier Isaac Hughes (55) killed by fall of stone.
23 September 1874	Tredegar	Collier Edward Lawrence (60) killed by premature shot. Died 13 October 1874.
12 March 1884	Bedwellty, Tredegar	Collier Theophilus Jones (40) killed by fall of stone. He ran out to the end of a stall to talk to a man who was doing some repairs, some coal fell and knocked out a pair of timbers and the deceased in trying to get out of the way was caught by the fall.
18 April 1884	Whitworth No. 2 Pit	Collier Emanuel Davies (30) killed by fall of coal. He was working the Fire Coal leaving the Yard Coal for a top and the collier next to him advised him to pull some of the top coal down, which he said he would. Nothing more was heard of him till, as he did not answer when called to go to dinner, he was searched for and found dead under a fall.
28 March 1888	Bedwellty No. 2 Level	Collier David Davies (37) killed by fall of roof from slips.
23 May 1906	Coronation, Cwmfelinfach	Collier Samuel Evans (37) killed.

Once again a sad reminder of the true price of coal.

Four

Beaufort to Crosskeys in the South Wales Coalfield

On Thursday 2 March 1871 an explosion claimed the lives of nineteen men and boys, three horses and a collier's dog at Victoria Colliery, Ebbw Vale. On that day, a thorough inspection of the coalface was undertaken by John Evans, the overman who gave the all clear. These findings were later confirmed by the company's mineral agent who had also been testing the workings with a naked flame. Naked flames were used to illuminate the roadways within the mine.

At approximately 4.00 p.m. a muffled roar came from deep within the workings. Fifty-year-old fireman Johnathen Price with his deputy made a descent to investigate. On their arrival at pit bottom, both men began an intensive search of the workings. Johnathen Price unfortunately took a wrong turning. The deputy took a different route and followed the roadway, down which the fresh air travelled. After a considerable time, and having received no communication from pit bottom, John Evans took it upon himself to be lowered into the darkness. Moving a short distance into the main headings he stumbled across the lifeless body of the fireman. Thinking that he was unconscious, he heroically carried his colleague to pit bottom. He was almost overcome by the choking fumes of afterdamp. There he came upon survivors, about twenty in number who had miraculously escaped both the explosion and the poisonous gas.

It was to be late evening before a mines' inspector arrived and signalled a fresh attempt to descend the pit. Having cleared the accumulation of foul gas, the search of the workings revealed that little damage had been sustained to the fabric of the mine. However, the cost to human life was high with four miners killed immediately while the remaining fifteen appeared to have succumbed to the suffocating carbonic acid vapour.

One of the first victims to be brought to the surface was fireman Johnathen Price. After making the initial descent into the gas filled mine, he had surrendered his life in a vain attempt to rescue those in peril below. The Price household was to suffer a double bereavement; among the dead was also a son, eighteen-year-old collier John Price. Yet, as so often happens, the victims of violent death are sometimes the most innocent. The true horror of the disaster was shown in the deaths of two youngsters, nineteen-year-old collier Samuel Cooke who had started work at the pit just a few weeks earlier so that he could better support his recently widowed mother and of twelve-year-old doorboy Joseph Harris. Five days later an inquest was held at which the jury decided to view each victim in turn. The inspector, Lionel Brough, giving evidence before the coroner, reported that a blower of gas had escaped at the coalface and been ignited by a naked flame. He further added that inadequate ventilation due to abnormal atmospheric conditions further aggravated the situation. It is interesting to note that while plans had been in existence for some time to improve the ventilation system, these had never been implemented. Remarkably the jury, after returning a verdict of accidental death, failed to mention the need for better ventilation.

Victoria No.1 Pit, Ebbw Fawr Valley, Monmouthshire in 1910. During the mid 1840s construction began on a new mine at Ebbw Vale, and was completed in 1846. Although comprising of only one shaft, both coal and iron ore were raised. Victoria was considered to be a safe pit, but within a year of its opening an explosion occurred. Despite this, the faith that the workings were free from hazard was not diminished. Naked flames, although thought to have been the cause of the blast, continued to be used. The disregard of this early warning was to have tragic consequences. On Thursday 2 March 1871, a second explosion claimed the lives of nineteen men and boys. Victoria No.1 Pit was abandoned in July 1915.

Opposite, above: Victoria No.5 Pit, Ebbw Fawr Valley, Monmouthshire in 1915. No.5 Pit was sunk in 1850 by Ebbw Vale Iron & Coal Co. On 25 July 1891 the accidents reports show that twenty-year-old labourer William Sutton was killed by a fall of roof in consequence of a sudden squeeze at a parting where he and others were loading timber into drams to send into the workings. Victoria No.5 Pit was abandoned in 1922.

Opposite, below: Victoria No.8 Pit, Ebbw Fawr Valley, Monmouthshire in 1965. Ebbw Vale Victoria No.8 Pit was sunk in the 1850s by Ebbw Vale Iron & Coal Co. On 26 May 1853 the accidents reports show that twenty-three-year-old collier John Gare was killed by a fall of coal. Victoria No.8 Pit was abandoned in 1922.

Marine Colliery, Cwm, Ebbw Fawr Valley, Monmouthshire in 1915. The mine was sunk in 1889 by the Ebbw Vale Steel, Iron & Coal Co. The NCB acquired the colliery on Vesting Day, 1 January 1947. Seams worked were the Old Coal, Big Vein, Lower Nine-Feet, Four-Feet and Five-Feet.

Opposite, above: Waunlwyd Colliery, Ebbw Fawr Valley, Monmouthshire during sinking in 1876. The mine was sunk in 1876 by Ebbw Vale Steel Iron & Coal Co. Shaft depth 816ft. On 1 January 1902 the accidents reports show that twenty-six-year-old haulier Thomas John Lewis was killed when his journey caught a stick of timber lying at the side of the road, causing it to strike and fracture his leg. He died on 6 February 1902.

Opposite, below: Waunlwyd Colliery in 1920. In 1955 with a manpower of 703 it produced an annual output of 184,000 tons; in 1958 with a manpower of 633 it produced an annual output of 130,000 tons and in 1961 with a manpower of 479 it produced an annual output of 118,000 tons Waunlwyd Colliery was closed on 30 December 1963 by the NCB.

N A N A

Above: Deep Pit shaft remains, Nantyglo, Ebbw Fach Valley, Monmouthshire in 1976. Deep Pit, Nantyglo, was closed in 1925. In 1921 coal production in the South Wales Coalfield ceased following a lockout at Welsh pits.

Opposite, above: Marine Colliery on St David's Day, 1 March 1927. On this day an explosion killed fifty-two men and boys. There were 1,4000 men employed at the colliery at this time but fortunately, on the day of the explosion, only the night shift were employed below ground. This consisted of 135 miners who worked the seams known as Old Coal and Black Vein. Many of the men from the lower seam had managed to reach safety shortly after the disaster had struck.

Opposite, below: Beynons Colliery, Blaina, Ebbw Fach Valley, Monmouthshire in 1974. The mine was sunk in 1922 by Lancaster's Steam Coal Collieries Ltd. Seams worked were the Lower Threequarter, Black Vein and Meadow Vein. Beynons Colliery was closed in consequence of an underground fire on 2 April 1975 by the NCB.

Thomas Brown of Blaina Iron Works, Ebbw Fach Valley, Monmouthshire and later first coal owner to exploit the Alled seam at Cwm Tillery in 1841.

Mining historian Harry Rogers with an oil burning lamp that was used on pit bottom. The lamp has two water tanks at the rear to cool the heat produced by the flame.

In 1888 South Wales Colliery Co. granted a lease of the pits to Lancaster Spier & Co. This lease was surrendered in 1891 and a new lease granted to Lancaster's Steam Coal Collieries Ltd, who in 1916 acquired all the assets of the South Wales Colliery Co. The Cwmtillery Pits were owned by Lancaster's Steam Coal Co. Ltd until Vesting Day, 1 January 1947.

Approximately 32 million tons of coal was raised at Cwmtillery between 1850 and 1950. The highest daily output was 2,085 tons on 25 May 1911; the highest weekly output 11,725 tons was also in May 1911.

During the 130 years of its life, methods of working had changed. Originally the pillar and stall method was used, later this was changed to longwall heading and stall and longwall Barry system. Cwmtillery was one of the first collieries in South Wales to use the Meco-Moore Power Loader. In 1977 the longest manriding installation in the South Wales Coalfield was commissioned to transport men and materials 9,300ft into the Garw seam. No.1 shaft depth 2,349ft to Old Coal seam; No.2 shaft depth 1,665ft to Black Vein; No.3 shaft depth 2,160ft to Old Coal seam.

Cwmtillery Colliery and Rose Heyworth Colliery merged with Abertillery New Mine on 21 November 1960.

Left: A coal cutter at Abertillery New Mine, Abertillery, Tillery Valley, Monmouthshire in the 1960s.

Roof bolting at Abertillery New Mine in 1982. Abertillery New Mine was a modern, high-output mine based on the linking of two neighbouring mines Cwmtillery and Rose Heyworth.

In 1956 a £3 million scheme involved the driving of a new 3,595ft drift mine, to integrate the two collieries and streamline coal handling. The unit incorporated a central washery, which also handled coal from neighbouring Blaenserchan Colliery, which was transported underground and up the Abertillery Drift.

In 1976 the workings of the mine spread beneath an area of around eight square miles, lying to the north and west of the Clydach Bridge and Greenland Faults which separate it from Blaenserchan and further north, Blaenavon.

In the centre of the mining area, the Cwmtillery Fault breaks the seams, with a 50-80ft drop from one side to the other. There were thirty-one miles of underground roadway in use, carrying more than eleven miles of high-speed belt conveyors.

Above, left: Abertillery New Mine General Manager, Raymond Maurice Young. Mr Young was born on 8 July 1929 and started work with John Painton & Partridge Jones Coal Co. at Six Bells Colliery in 1945. His first job was a collier's assistant, then he became a measuring clerk, a fireman in 1954, an overman in 1955, a full-time student from 1956-1959, an overman at Rose Heyworth Colliery from 1959-1960, an undermanager at Rose Heyworth Colliery from 1960-1963, a manager at Blaenserchan Colliery from 1963-1964, a manager at Celynen South Colliery from 1964-1966, a manager at Penrikyber Colliery from 1966-1975, a general manager at Abertillery New Mine (Cwmtillery Colliery and Rose Heyworth Colliery) from 1975-1983, a manager at Penallta Colliery from 1983-1986, a site engineer at Nantgarw Colliery from 1986-1987 (to fill shafts and clear site), and finally he retired in 1987, aged fifty-eight, having served forty-two years in the coal industry.

Above, right: Miners' drinking water Jack and Food Box (Tommy Box).

In the early days of mining it was thought that the seams ceased at the Old Coal level and the rock band lying beneath this seam was named 'Farewell Rock'. Later, however, exploratory bores through this rock revealed the rich Garw seam. Coal was also mined from the Black Vein seam at a depth of 550ft.

Developments of Abertillery include a 360ft underground tunnel, driven through to nearby Blaenserchan Colliery to streamline coal transportation between the two collieries. £100,000 had also been spent on a new manriding installation, at 9,842ft the longest in the South Wales Coalfield.

In 1976 with a manpower of 987 (underground) and 88 (washery) it produced an annual saleable output of 143,306 tons; it produced an average weekly saleable output of 5,314 tons; average output per man/shift at the coalface 5 tons 3cwt; average output per man/shift overall 1 ton 6cwt; deepest working level 1,640ft; number of coal faces that were working, 4; seams working, Lower Nine-Feet & Upper Bute (Lower Black) and Garw; drift length 3,595ft; gradient 1 in 5; stocking capacity on pit surface 40,000 tons; average weekly washery throughput 11,711 tons; types of coal coking; markets (principally) steel industry; fan capacity, cubic feet per minute 225,000; average maximum demand of electrical power 3,150kW; total capital value of plant and machinery in use was £1.1 million. Abertillery New Mine was closed on 9 October 1985 by British Coal.

Gray Colliery, Abertillery, Tillery Valley, Monmouthshire in 1909. The mine was opened in 1849 by Powell's Tillery Steam Coal Co. Gray Colliery was closed in 1938. Adult life for the children of a mining family could begin at a very early age. Boys and girls as young as six years old worked underground in the early years of the coal industry. Many of them were employed as doorkeepers. Their job was to open and shut doors, which cut off sections of the workings underground and which helped to control the ventilation of the mine.

Edward Prosser, Deputy Manager, in 1890 when the Pit opened. Vivian Colliery was closed on 19 July 1958 by the NCB.

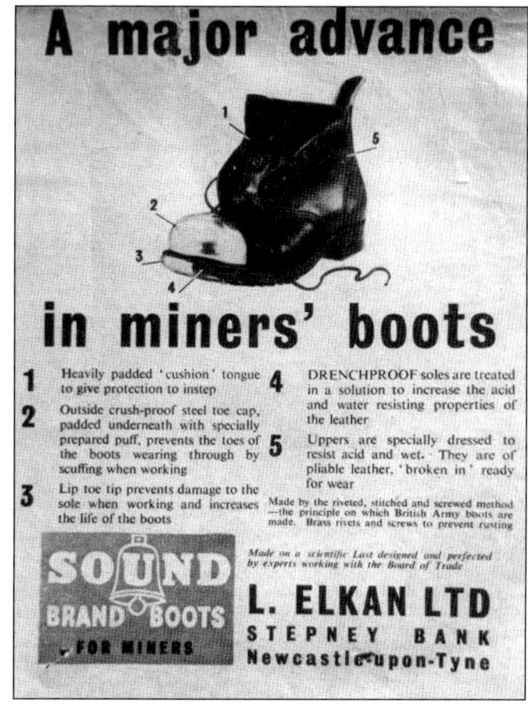

A colliery newspaper coal advertisement for Pit Safety Boots in December 1947. In 1947 the *Coal* newspaper was 4*d* (approximately 2p).

Sinking Arrael Griffin, also known as Lancaster's Colliery and Six Bells Colliery, near Six Bells, Ebbw Fach Valley, Monmouthshire in 1890. In 1890, the year Queen Victoria opened the great Forth Bridge, John Lancaster & Co. were extending their activities south from the Ebbw Fach, sinking two shafts at Six Bells, which were to become the Arrael Griffin Colliery. The Colliery reached its peak manpower figures in 1914, when it employed 2,857 miners and, by the early 1930s, annual coal output stood at more than 600,000 tons. In 1936, ownership passed to Partridge, Jones & Paton, already owners of a number of other collieries in south-western Monmouthshire, until Nationalisation in 1947. Since then, when the mine began to be known as Six Bells Colliery, more than £400,000 was spent on surface modernisation, including electrification of the shaft winding in the early 1960s.

On 28 June 1960 an explosion caused the death of forty-five miners, the youngest being eighteen years old and the eldest sixty years old.

Following the publication in 1973 of *Plan for Coal* a massive capital investment programme was launched by the NCB. This colliery was one of those involved in the programme. Underground streamlining took place as part of a development linking the pit with neighbouring Marine Colliery, in the Ebbw Fawr Valley (Marine Colliery closed in March 1989).

Six Bells coal travelled directly underground to Marine Colliery where it was wound, washed and graded before despatch to its eventual markets. The mining programme at Six Bells worked an area of around four square miles, taking in several major geological faults running roughly north to south and varying the level of coal seams by between 70ft and 130ft. One of these complicated the mining operation by splitting the 'take' and, working across this had called for all the traditional 'know how' of Welsh mining engineers.

In 1976 there were almost seven miles of underground roadway in the mine and around two and a half miles of high speed belt conveyors in daily use. Coal was being worked from the Garw seam at a maximum working depth of around 1,800ft, though, during its lifetime, the pit had also worked the Big Vein, Elled, Meadow Vein, Black Vein, Old Coal and Threequarter seams. The 638 miners at Six Bells produced an average 109,568 tons a year, mainly of prime coking coal and top-quality steam coal for industry.

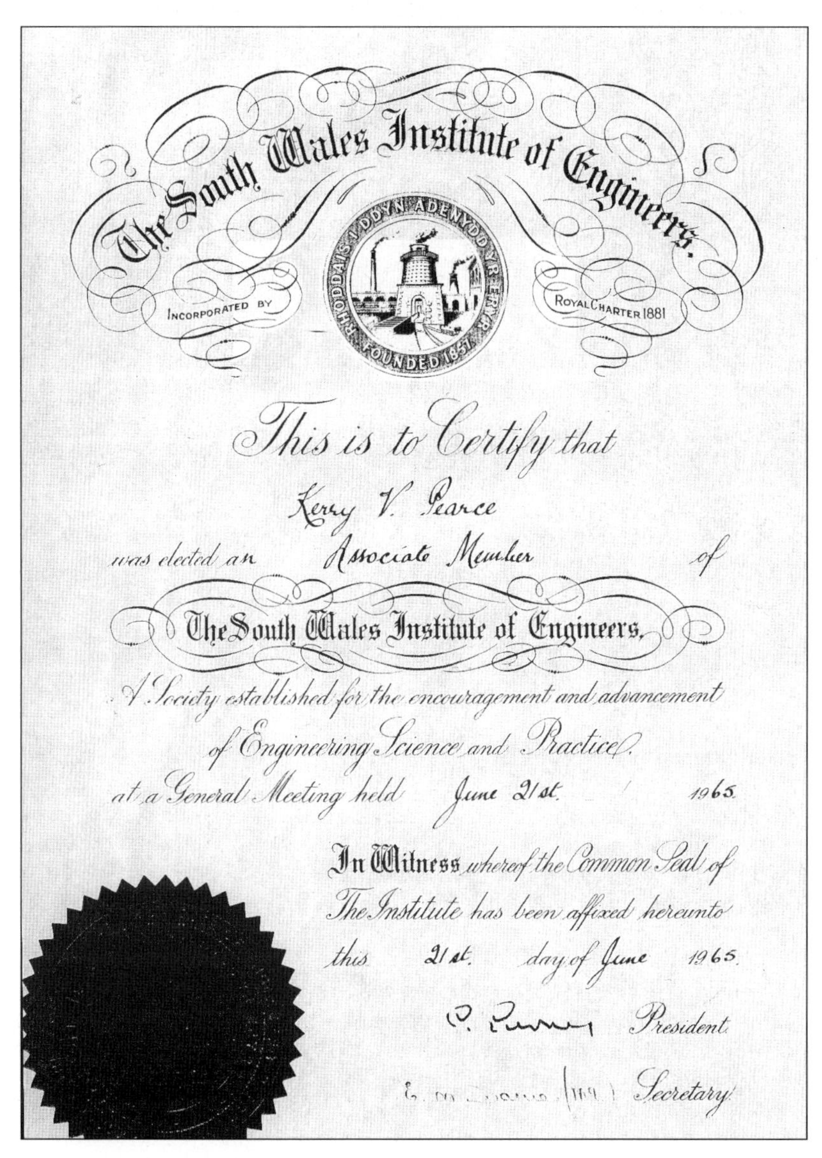

The South Wales Institute of Engineers.

Incorporated by Royal Charter 1881

Rhoddais i ddyn adenydd yr eryr · Founded 1857

This is to Certify that

Kerry V. Pearce

was elected an Associate Member of

The South Wales Institute of Engineers,

A Society established for the encouragement and advancement of Engineering Science and Practice

at a General Meeting held June 21st 1965

In Witness whereof the Common Seal of The Institute has been affixed hereunto this 21st day of June 1965.

President

Secretary

South Wales Institute of Engineers Certificate, presented to Kerry Verdun Pearce.

THE INSTITUTION OF MINING ENGINEERS

FOUNDED 1st JULY 1889

Incorporated by Royal Charter 9th February 1915

This is to certify that

Kerry Verdun Pearce

was elected an Associate Member (Group A) of

THE INSTITUTION OF MINING ENGINEERS

on the Twenty-fourth day of June 1965

Witness our hands and Seal at London this

Thirtieth day of July 1965

President

Secretary

The Institute of Mining Engineers Certificate presented to Kerry Verdun Pearce. Mr Pearce was born on 23 September 1938 and started work at Fernhill Colliery in 1955. His first job was as a collier's assistant; then from 1955-1960 he went to day school one day a week at Rhondda College and became a measuring clerk; from 1960-1963 he took a Mining Sandwich Diploma Course at Glamorgan College of Technology, Treforest; he became a fireman at Parc Colliery from 1963-1965 and in 1965 an Associate Member of the Institute of Engineers and Associate Member of the South Wales Institute of Engineers; from 1965-1968 a fireman and then promoted to overman Fernhill Colliery; from 1968-1980 an undermanager at Mardy Colliery; a manager at Tower Colliery from 1980-1984; a manager at Six Bells Colliery and Blaenserchan Colliery from 1984-1988; a manager at Marine Colliery 1988-1989 and finally in 1989 he retired, aged fifty-one, having served thiry-five years in the coal industry.

In 1976 with a manpower of 638, Six Bells Colliery produced an annual saleable output of 109,568 tons; it produced an average weekly saleable output of 1,771 tons; average output per man/shift at the coalface 3 tons 5cwt; average output per man/shift overall 14cwt; deepest working level 1,800ft; number of coal faces that were working, 2; No.1 shaft depth 813ft, diameter 18ft; No.2 shaft depth 1,154ft, diameter 20ft; manwinding capacity per cage wind, 24; all coal was wound at Marine Colliery; winding engines horsepower 900/1,950; types of coal coking; markets steel; fan capacity, cubic feet per minute 180,000; average maximum demand of electrical power 6,104kW; total capital value of plant and machinery in use was £273,801; estimated workable coal reserves 5.4 million tons.

Six Bells Colliery Coal Seams:

Depth Feet	*Standard Name*	*Local Name*
770	Two-Feet-Nine	Elled
800	Upper Four-Feet & Four-Feet	Big Vein
830	Six-Feet	Threequarter
903	Upper Nine-Feet	Upper Black Vein
913	Lower Nine-Feet	Lower Black Vein
974	Yard & Seven-Feet	Meadow Vein
1,029	Five-Feet & Gellideg	Old Coal
1,109	Garw	Garw

Six Bells Colliery was closed in March 1989 by British Coal.

This poor quality image dating from 1905 shows Hafod Fan Level, near Six Bells, Ebbw Fach Valley, Monmouthshire. The mine was opened in 1870. On 15 March 1898 the accidents reports show that twenty-seven-year-old clerk William Arthur Harrington was fatally injured. Hafod Fan Level was abandoned on 31 December 1908.

Hafodyrynys Colliery (old), near Hafodyrynys, Glyn Valley, Monmouthshire in 1940. The mine was sunk in 1878 by E. Jones. Shaft depth 908ft. From 1914 to Vesting Day, 1 January 1947, the mine was owned by Crumlin Valleys Colliery Co. Seams worked were the Meadow Vein and Big Vein. Old Hafodyrynys Colliery was closed on 7 July 1966 by the NCB.

Glyn Pit remains and Beam pumping engine in 1982.

Glyn Pit remains in 1982. On the left in the photograph is a 1956 Triumph 500cc Speed Twin motor bike belonging to Colin Evans. Following closure in 1932 Glyn Pit was used for pumping minewater from Old Hafodyrynys Colliery which was closed on 7 July 1966 by the NCB.

A lighthouse type battery-operated safety hand lamp showing the battery (left), locking base (centre) and lamp section. The Concordia hand lamp weighed 9½lbs; the average weight of hand lamps used was 7lbs. By the 1930s electric hand lamps became common in the coalmines, to be replaced by helmet-mounted electric cap lamps following Nationalisation. The advent of battery-powered electric lights led to a dramatic drop in cases of miners' nystagmus, an eye disease caused by working in poor light.

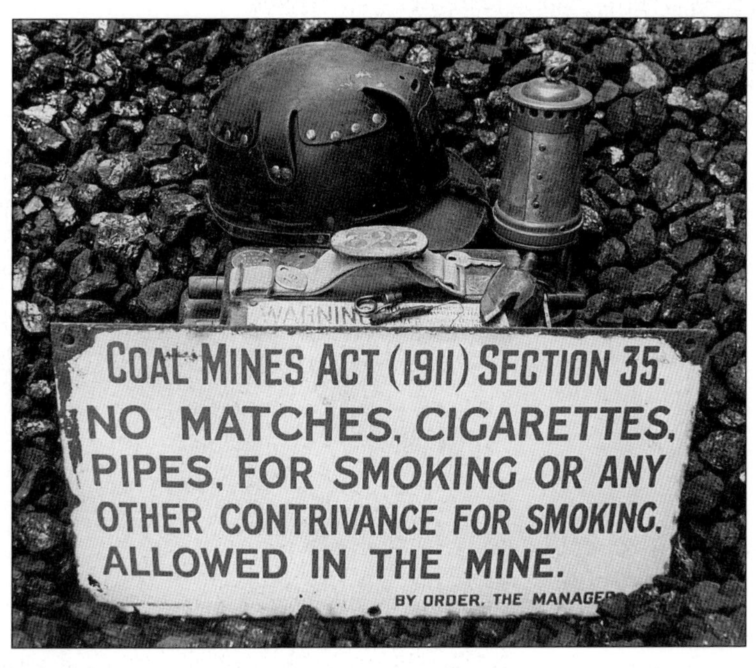

This photograph includes: miners safety helmet, shotsman's safety lamp (a lamp in which dangerous contact between the external atmosphere and a naked flame is prevented), safety lamp relighting key, lamp check, shot-firing battery (a shotsman is a qualified official who fires shot holes in a district), detonator, key to morphia container, coal boring bit and a Coal Mines Act (1911) sign used at collieries.

N.C.B.
684
HAFODYRYNYS

A colliery newspaper advertisement for Climax Shot-Hole Drills in 1947. 'In 20 seconds this Climax Shot-Hole Drill will make a 3ft 9in hole. It takes only 2 minutes to drill a round of holes in a coalheading instead of the 10 minutes required with an ordinary percussive drill. Machines such as this are helping to win the Battle for Fuel! For details of pneumatic and electric drills write today to: Climax Rock Drill & Engineering Works Ltd'.

Opposite, above: Hafodyrynys Lamp Check, Glyn Valley, Monmouthshire. Between 1954 and 1959 the NCB developed the New Mine at a cost of £5½ million, incorporating Hafodyrynys New Mine and Glyn Tillery Drifts (closed 23 December 1975 by the NCB) and linking it to Tirpentwys Colliery (closed 29 November 1969) and later Blaenserchan Colliery (closed 28 August 1985).

Opposite, below: Treppaner coal cutter, Hafodyrynys New Mine in 1963.

Llanhilleth Colliery,

JUNE 15th, 1901.

APOLOGY.

I, the undersigned **ALBERT COLES**, of Preston Street, Abertillery, hereby express my regret that on June 13th, 1901, I committed a breach of Special Rule Number 91, established under the Coal Mines Regulation Act, 1887, by having in my possession in the Mine belonging to Messrs. Partridge, Jones & Co., Ltd., a Pipe.

And I hereby tender my sincere Apology to the Owners, the Officials, and my Fellow Workmen for so doing.

I further agree to pay the sum of Ten Shillings towards the Library Fund as proof of my penitence, and to pay the cost of printing this Apology.

I also undertake, in the future, to faithfully observe and support to the best of my ability, the Rules and Regulations of the Coal Mines Regulation Act, and the Special Rules established thereunder.

I also agree that a copy of this Apology shall be posted up at the Mine as a Warning to others.

Witness my Signature this 20th day of **JUNE, 1901,**

ALBERT COLES.

Witness to the above Signature **DAVID DAVIES.**

Llanhilleth Colliery Apology, 15 June 1901. The Apology reads:

I, the undersigned Albert Coles of Preston Street, Abertillery hereby express my regret that on 13 June 1901, I committed a breach of Special Rule Number 91, established under the Coal Mines Regulation Act, 1887, by having in my possession in the Mine belonging to Messrs Partridge, Jones & Co. Ltd a Pipe. And I hereby tender my sincere Apology to the Owners, the Officials, and my Fellow Workmen for so doing. I further agree to pay the sum of Ten Shillings [50p] towards the Library Fund as proof of my penitence, and to pay the cost of printing this Apology. I also undertake, in the future, to faithfully observe and support to the best of my ability, the rules and Regulations of the Coal Mines Regulation Act, and the Special Rules established thereunder. I also agree that a copy of this Apology shall be posted up at the Mine as a Warning to others. Witness my Signature this 20th day of June, 1901, Albert Coles. Witness to the above Signature David Davies.

Blaencuffin Level, Llanhilleth, Ebbw Fawr Valley, Monmouthshire in 1970. In 1966 this privately owned mine belonged to W.J. Desmond, Blaencuffin Farm, Llanhilleth and worked the Mynyddislwyn seam under the Inspectorate in the former South-Western Division. The miners with his working companion are preparing to take an empty dram into the level. Blaencuffin Level was abandoned on 21 April 2001.

Opposite, above: Aberbeeg Colliery, Ebbw Fawr Valley, Monmouthshire in 1901. The mine was opened in 1860 by Budd & Co. On 2 November 1906 the accidents records show that twenty-one-year-old collier William Smith was fatally injured. Aberbeeg Colliery was closed on 31 December 1965 by the NCB. In 1931 a coal dispute in the South Wales Coalfield put 140,000 men out of work.

Crumlin Navigation Colliery, Crumlin, Ebbw Fawr Valley, Monmouthshire in 1965. The mine was opened in 1911 by Partridge Jones & Co. Shaft depth 1,536ft. Seams worked were the Old Coal, Threequarter, Meadow Vein and Black Vein. On 1 January 1947 the coal industry was Nationalised.

Opposite, below: Celynen Colliery, Newbridge, Ebbw Fawr Valley, Monmouthshire in 1975. Left to right: Graig Fawr Pit, South Pit and North Pit. Celynen North Colliery was sunk by the Newport Abercarn Co. The mine produced prime coking coal for Ebbw Vale and Llanwern Steelworks. Seams worked were the Meadow Vein, Upper Threequarter, Big Vein and Black Vein.

Above: Crumlin Navigation Colliery single drum compressed air (blast) haulage engine in 1965. Haulage engines were driven by blast, steam or electrical type of fixed engine, which was used on the surface and underground. Underground it was used for taking supplies into a district, for the face and returning with a full journey of coal. Crumlin Navigation Colliery was closed on 16 September 1967 by the NCB.

Celynen North Colliery underground dump in 1960. In 1982 the mine was linked to Oakdale Colliery and coal was diverted to the Oakdale pit bottom. In 1985 the miners were transferred to Oakdale Colliery Unit. Celynen North Colliery was closed 31 March 1985 by the NCB.

Underground in the 1970s. Front right is Tom Jones (Bwllfa). Seams worked in 1967 were Old Coal and Big Vein. Celynen South Colliery was closed on 5 September 1985 by the NCB. From Nationalisation to 1956 a total of nearly £25 million had been spent on colliery-associated activities in the Coalfield. These activities included boring, the erection of pithead baths, canteens, medical centres, central washeries, experimental equipment, etc.

Right: Abercarn, Prince of Wales Colliery lamp check, Abercarn, Ebbw Fawr Valley, Monmouthshire. The mine was opened in 1836. On 11 September 1878 an explosion killed 268 men and boys and 57 horses. Twenty-seven years after the incident, miners found a complete skeleton in ragged clothing and boots. At the inquest it was held that he was one of the victims and unidentifiable. Abercarn, Prince of Wales Colliery was abandoned in 1922.

Opposite, below: Celynen South No.1 and No.3 Colliery, Abercarn, Ebbw Fawr Valley, Monmouthshire in 1971. The mine was sunk in 1873 by Newport Abercarn Black Vein Steam Coal Co. On 25 May 1888 the accidents records show that twenty-seven-year-old collier James Norwood was fatally injured. In 1966 the Celynen South manager was R.M. Young (7,764 First Class) and the undermanager was W.I. West (6,554 Second Class).

Cwmcarn Colliery, Cwmcarn, Ebbw Fawr Valley, Monmouthshire in 1935. The downcast shaft (left) was sunk in 1876 and the upcast shaft (right) was sunk in 1912 by the Ebbw Vale Steel Iron & Coal Co. In 1966 the Cwmcarn South manager was A. Radford (5,889 First Class) and the undermanager was E.G. James (6,115 Second Class). Cwmcarn Colliery was closed on 30 November 1968 by the NCB.

Risca Colliery, near Crosskeys, Ebbw Fawr Valley, Monmouthshire in 1960. The mine was sunk in 1872 by the London and South Wales Colliery Co. Shaft depth 1,009ft. Seams worked were the Yard/Seven Feet, Black Vein, Red Vein, Four-Feet and Big Vein. Risca Colliery was closed on 9 July 1966 by the NCB.

In Everlasting Memory to the Miners who lost their lives from Beaufort to Crosskeys

Date	Mine	Lives Lost
14 January 1846	Black Vein, Risca	Twenty-four miners killed.
December 1847	Nantyglo, Ebbw Vale	Eight miners killed.
12 March 1853	Risca, near Crosskeys	Ten miners killed. Those who died were: collier Aaron Brian (22), collier Joseph Bryant (24), collier Rees Davies (24), doorboy George Purnell (11), doorboy Moses Moore (12), collier Samuel Dack (12) killed by suffocation after explosion of firedamp, collier Solomon Jenkins (20) burnt and suffocated, collier Thomas Davies (40), collier John Williams (44), haulier George Phillips (19) burnt by explosion of firedamp.
30 November 1853	No. 15 Pit, Ebbw Vale	Collier Thomas Davis (48) and collier James Morgan (25) killed by an explosion of firedamp, want of ventilation and discipline.
19 February 1868	Tunnel Pit, Ebbw Vale	Underground stoker William James (24) killed when a terrific gale on the mountain on that day suspended the ventilation for a short time, in consequence of there being scarcely any stack on the upcast shaft, then the products of the combustion of the underground boiler fire backed and overpowered him.
6 June 1868	Black Vein, Risca	Collier H. Kingston (43) killed on the slope in consequence of fracture of draw-plate. There were plenty of safety stalls, but nevertheless he got overwhelmed by the rushing drams. He had that moment passed a safety hole, but no doubt his presence of mind was gone.
15 June 1870	Beaufort, Ebbw Vale	Miner William Harford (70) killed.
15 July 1880	Risca, near Crosskeys	Hundred and Nineteen killed by an explosion.
21 January 1884	Celynen, Ebbw Vale	Timberman John Abbot (33) killed by fall of stone. He was ripping top and in barring out some timber when it fell unexpectedly.
24 June 1884	Quarry Pit, Ebbw Vale	Collier Henry Garland (23). Seven men were coming out together in an empty dram. They noticed the top working over them and all got out and escaped except one, who was in the bottom of the dram and he was caught by the stone and killed.
8 August 1884	Aberbeeg, Ebbw Vale	Labourer John Prosser (63) killed after weighing a wagon of small coal he was lowering it down the siding, which was his customary work. The wagon ran over him.

Once again a sad reminder of the true price of coal.

Five

Blaenavon to Newport Docks in the South Wales Coalfield

Coal getting and transport underground was highly mechanised for many years. The same applied to cutting the coal in the seam, but the mechanisation of most of the other coalface operations was difficult if only because of the great variety of seam conditions. The proportion of coal cut by machine increased from seventy-five per cent in 1947 to about eighty-six per cent in 1955 and could hardly be raised much higher. But both shot-firing and loading in normal mining practice remained manual operations. Mechanisation by a combined cutting and loading process was, therefore, a main development objective. This was generally called 'power-loading'. Only about five million tons of coal was power-loaded in 1947 and the only types of machines available were of limited application. The period from 1955 had seen the development of many new machines and about a dozen different types were available, while a number of others were in various stages of development. The NCB were, therefore, in a position to apply power-loading on a much wider scale than ever before. In July 1955, they launched a far-reaching programme to establish power-loading units throughout all coalfields, wherever physical conditions permitted. The saleable output obtained by machines of these kinds increased from 7½ million tons in 1950 to 23 million tons in 1955, and it was hoped to double this figure within three years.

The first task of the Board's new Central Engineering Establishment was the development of improved coalface machinery, and the research problems arising from this work had been given the highest priority at their Mining Research Establishment. The NCB's Mechanisation Branch had been strengthened and expert teams of experienced pitmen were being set up in all areas to install power-loading machines and to train men in their efficient use. Another aim was the development of a fully mechanised support system and experimental equipment was undergoing trials underground. Much use was already being made of adjustable props. In 1955 they were over ten per cent of the total. But a fully mechanised support system would go much further. The engineers had a vision of the future in which the coalface moves forward much more rapidly than before, with the coal being cut and loaded mechanically, and with roof-support equipment following under its own power.

Coal transport was fully discussed in *Plan for Coal*. Locomotives, belt conveyors and rope haulage remained the primary means. The number of locomotives in use underground increased from 213 in 1949 to 300 in 1951 and to nearly 700 in 1955, and locomotive haulage was extended wherever roads were suitable. Some 15,000 large minecars (over 30cwt) had been installed.

There were novel means of transport under development although they were still economically speculative. In particular, two experiments in the hydraulic transport of coal had been carried out in which the coal is mixed with water and is then pumped in pipes to the surface.

Young scholars at Blaenavon Colliery known as Big Pit (Pwll Mawr), Blaenavon, Afon Lwyd Valley, Monmouthshire in 1905. Kearsley's Pit was sunk in 1860 and later became Big Pit (Pwll Mawr).

On 21 October 1872 the accidents reports show that thirteen-year-old doorboy David Henry Jones was fatally injured by a fall of roof and on 24 August 1889 the accidents reports also show that sixteen-year-old collier John O. Jones was fatally injured by a fall of roof. The place, it is alleged, was well timbered at the time.

Around 1878-79 Kearsley's Pit, which had been used as a ventilation shaft for the Forge Level and Pit, was deepened from the Threequarter seam depth of 128ft to the Old Coal at 293ft. The new colliery appeared in the Mines Inspectors' List of Mines as Big Pit in 1880 and by 1895 it was raising 5,000 tons of coal per week with Dodd's Slope.

In the 1890s Big Pit worked five of the seams by the longwall system; the sixth, the Elled seam, was worked on the pillar and stall method, horses being employed in hauling from the face workings to the bottom of the main road. From this point the drams were hauled by steam engine to the pit bottom. There were four haulage engines in Big Pit supplied with steam from four Cornish boilers. One engine in the Old Coal had two 18in. cylinders, and worked a haulage plane over 3,600ft in length, with gradient of 1 in 12.

The original winding engine was constructed by Fowler Co. of Leeds. It had two 26in horizontal cylinders and drums of flat ropes and could raise two drams in each cage.

Thirteen Cornish and egg-ended boilers supplied steam to the winding engine, at the engines on Dodd's and Forge Slopes. The upcast shaft for Big Pit was the Coity Pit which had a Waddle fan. In 1950 the pit was wound by an 1880 Fowler Steam winding engine with a flat rope.

Until it closed on 4 April 1980, Blaenavon had some of the oldest workings in South Wales. Its drainage system was unique: it incorporated the Forge Slope, Wood's Level and Kear's Slope, all of them dating back to 1830 and before, the Engine Pit Level *c.* 1810 was also retained as an emergency exit.

Today Big Pit is the National Mining Museum of Wales and you can experience a free guided tour 300ft below the Welsh hillside by ex-miners, who will show you the mining industry as it really was. Blaenavon – a World Heritage Site created by coal mining and iron working, is one of the best examples in the world.

The first recorded working of coal in the Blaenavon area was in 1775, when William Tanner and Mary Gunter were granted a lease to work coal. The early workings were patch works and small levels situated on and around the extensive outcrop; many of these early sites still survive today to the north and west of Blaenavon. It is probable that the first shaft to be sunk was the Engine Pit, which was sunk as a pumping pit. The sinking date is not known but would have been around 1800 to 1810; the Engine Pit Level would also have been driven about the same time. The next major development would have been the sinking of the Cinder Pits, which are shown on the 1819 plan of Blaenavon, as intended Engine Pits for the deep work. The Cinder Pits were the scene of Blaenavon's worst mining disaster, on 28 November 1838 when torrential rain, after a heavy snowfall, ran off the hillsides and poured down the shafts of the Cinder Pits, drowning fourteen men and two women: nine-year-old Elizabeth Havard and sixteen-year-old Mary Hale.

Big Pit Tour Guides in July 2002. The photograph includes: Martyn 'Blob' Davies (former pipe fitter), William 'Billy Rip' Richings (former deputy), Glyn 'Lord Hydrant' Hallett (fire officer), Mike Keenan (banksman), Gerald 'Blue Bells' Andrews (former overman), Gavin 'Trogg' Rogers (pitman), Alan 'Ginger' Jones (former fitter).

Big Pit (Blaenavon Colliery) coal seams:

Depth	*Standard Name*	*Local Name*
76ft	Two Feet Nine	Elled
100ft	Upper Four-Feet	Big Vein
127ft	Upper Six-Feet	Threequarter
196ft	Upper Nine-Feet	Horn or Top Rock
215ft	Nine-Feet	Bottom Rock
224ft	Lower Nine-Feet & Upper Bute	Black
295ft	Yard	Meadow
311ft	Seven-Feet	Yard
370ft	Lower Five-Feet & Gellideg	Old Coal
442ft	Garw	Garw

Within the operation was 7.8 miles of underground roadways and almost 1½ miles of high-speed belt conveyors were in daily use speeding the coal directly up the 1,860ft, 1 in 6 to 1 in 4 New Mine drift to the surface. From there they were taken onward to the neighbouring washery for the eventual market, the giant British Steel Corporation plant at Llanwern, Newport.

Blaenavon New Mine, Blaenavon, Afon Lwyd Valley, Monmouthshire in 1980. In 1973, coal winding at Big Pit ceased. A new drift had been driven near the washery enabling all coal to be raised, washed and blended on one site. From the late 1960s, all production was concentrated in the Garw seam. The maximum section of coal of the Garw was 2ft 6in, its minimum 2ft 2in. Blaenavon New Mine and Big Pit complex was closed on 4 April 1980 by the NCB.

The Water Run between Coity Colliery and Big Pit, Blaenavon, Afon Lwyd Valley, Monmouthshire in 1965. It was driven in 1801 and used as a water drainage level. In 1913 at Universal Colliery a sad chapter in the history of the South Wales Coalfield came to a close. It was a chapter marked with the tragic loss of 439 men and boys and the immeasurable grief of an entire valley community and in 1966 the last major disaster in the South Wales Coalfield occurred which, broke the heart of a nation, Aberfan. The whole world was shocked by a tragic tip slide; with its appalling toll of 144, including 114 school children, two children not school age and twenty-eight adults, six of them teachers.

Kay's Slope, Blaenavon, Afon Lwyd Valley, Monmouthshire in 1966. A slope is a tunnel driven at a gradient through strata and seams into the required seam, normally at a downward inclination to connect underground workings with the surface, for the transportation of coal and supplies (on rails for drams). The coal was then mined without having to sink a shaft. Kay's Slope was opened in 1865.

Opposite, below: River Arch water drainage level for Big Pit, Blaenavon, Afon Lwyd Valley, Monmouthshire in 1965. River Arch water drainage level is still in use today and is an emergency second way out to the surface for Big Pit. A selection of Levels and Collieries working before 1835: Bridge Level 1782; Aaron Brutes Level *c.* 1800; Engine Pit and Engine Pit Level 1800 to 1810; Black Pin Level 1812; Dick Kear's Slope (New Slope) *c.* 1820; Forge Slope 1830.

Kay's Slope colliers holing in 1910. On the left in the photograph is a burning candle and on the right the collier is holing (making the horizontal cut in the seam) the coal in a band of inferior coal or shale part way up the seam. More usually holing was done at floor level, but could be done at roof level in some seams and was called over cutting. Whichever part of the seam was holed the purpose was to free the coal in as large lumps as possible; there was little or no market for small coal and the payment was at a much lower rate for the collier. In many collieries all small coal was simply thrown into the gob. Kay's Slope was closed on 1 January 1961 by the NCB.

Garn Slope, near Blaenavon, Afon Lwyd Valley, Monmouthshire in 1975. The mine was opened in 1925. Garn Slope merged with Kays Slope in 1958 and Big Pit in 1960. Garn Slope was closed on 1 January 1961 by the NCB.

Right: Front Holt miner Roy Seldon in 1964.

Left: Front Holt miner Dai 'Elvis' Owen in 1964. Roy and Dai removed stone from above the seam to create a higher heading at the entry to the coalface at the gate road and erected steel rings which were then lagged with timber. A conveyor in the gate road carries the coal outbye. The supply road supplies the face and district with timber, props, rings, rails and sleepers.

MONMOUTHSHIRE EDUCATION AUTHORITY

Approved Under

COAL MINES ACT. 1911. No 2951.

Certificate of Qualification of Fireman, Examiner, or Deputy under Section 15(1)(b).

This is to certify that Trevor George Williams,

residing at Coronation Road, Cwmbran,

has been duly examined and has satisfied the Examiners

(A) That he is able to make accurate tests (so far as practicable with a safety lamp) for inflammable gas;

(B) That he is able to measure the quantity of air in an air current;

(C) That his hearing is such as to enable him to carry out efficiently the duties of fireman, examiner or deputy.

Signed on behalf of the Authority. Dated 13th June, 1914

Sam M Jones Chairman of the Higher Education Sub-Committee.

Examiner.

County Medical Officer.

Director of Higher Education.

Monmouthshire Education Authority, approved under Coal Mines Act 1911. No. 2951. Certificate of Qualification of Fireman Examiner, or Deputy under Section 15(1) (b). 'Trevor George Williams residing at Coronation Road, Cwmbran, has been duly examined and has satisfied the Examiners.'

Garn Pit remains in 1965. The mine ceased coal production in 1890. Garn Pit was abandoned in 1922.

I'm trying not to cry see dad, I want to be brave like you, I know I must come here every day so I can earn money too. But one little candle don't show much light and it's hard to breathe here as well; when the preacher was talking in chapel last night, was this what he meant by 'Hell'?

Opposite, below: Dram of coal discovered at opencast mining, near Blaenavon, Afon Lwyd Valley, Monmouthshire in 1973. The most northerly Colliery in the South Wales Coalfield was Beaufort Colliery, Monmouthshire; the most southerly colliery was Cardiff Navigation Colliery, Lanelay, Glamorganshire; the most westerly colliery was Trefane Cliff Colliery, near Newgale, Pembrokeshire and the most easterly colliery in the South Wales Coalfield was Elled Colliery, Blaenavon, Monmouthshire.

Milfran Colliery, nr Blaenavon, Afon Lwyd Valley, Monmouthshire. The mine began sinking in 1885 by Blaenavon Coal & Iron Co. On 10 July 1929 an explosion killed cutterman Evan Howells (40), collier assistant Clifford Edmunds (20), collier Ernest Holder (57), collier David Ricketts (35), collier David John Parry (37), collier assistant Evan James Willaims (20), collier assistant Albert James Williams (21), and collier assisitant Ernest Southcott (20) who died from toxaemia 18 July 1920.

Above: Construction of an underground haulage engine house in 1933. A haulage engine is a steam, compressed air, or electrical type of fixed engine, surface or below ground. Used for the taking into a district, supplies for the face and returning with a full journey of coal. Milfraen Colliery was closed in 1935.

Opposite, above: A coalmining family in the 1930s. Jack and Catherine's three sons left school at fourteen and followed their father into the coalmining industry, and became heading men. Left to right: Brian 'Big Bird' Evans, Jack Evans (block layer), Cyril 'Little Bird' Evans, Catherine Anne Evans, Bryn Evans. The Coal Industry Social Welfare Organisation organised many competitions for miners and their families: athletics, bowls, brass band, coal queen, darts, course, fly and sea angling, football and rugby, golf, light opera, holiday schemes, carnivals, grants etc.

Opposite, below: Blaenserchan Colliery, near Pontypool, Afon Lwyd Valley, Monmouthshire in 1905. The mine was opened in 1890 by Partridge Jones & Co. Seams worked were the Meadow Vein, Old Coal, Black Vein, Elled and Big Vein. Blaenserchan Colliery, the last deep mine in the Afon Lwyd Valley, was closed on 28 August 1985 by the NCB.

Tirpentwys Colliery pit bank, near Pontypool, Afon Lwyd Valley, Monmouthshire in 1960. The mine was sunk in 1878 by Darby & Morris. It was later purchased by the Tirpentwys Colliery Co. Shaft depth 1,327ft. Seams worked were the Five-Feet/Gellideg, Bute, Big Vein and Lower Six-Feet. In 1959 production was diverted to Hafodyrynys New Mine.

Opposite, above: Tirpentwys Colliery Pit Bottom in 1909. In 1956 with a manpower of 893 it produced an annual output of 198,000 tons. Tirpentwys Colliery was closed on 29 November 1969 by the NCB. 'And do they have rats in hell our dad? Cos I just saw a big one run by, I'm scared of it dad, and everything here, but I'm trying hard not to cry. So, how long have we got to stay here our dad? How much coal have we got to get? Cos I'd rather be out and playing far away from this dark and wet.' H. Rogers (Granch).

Opposite, below: Llanerch Colliery, near Pontypool, Afon Lwyd Valley, Monmouthshire in 1964. The mine was opened in 1858 by the Ebbw Vale Steel, Iron & Coal Co. Seams worked were the Big Vein and Threequarter. On 6 February 1890 at approximately 8.30 a.m. 176 men and boys were killed including undermanager Edward Jones, by an explosion of firedamp at naked lights in the No.4 heading off Cooke's slope. The miners had been at work in their places for just under two and a half hours, the youngest being twelve-years old and the eldest sixty-four-years old.

Above: Underground in 1946. Llanerch Colliery was closed on 10 September1947 by the NCB. In 1984 the miners went on strike against pit closures. The 1884-85 strike was a fight to save jobs. However, perhaps many more view the loss of jobs as a small price to pay for an end to the terrible toll of human life, the suffering and the desecration of a once beautiful landscape, which were hallmarks of an era when coal was king. There was always tremendous courage and staunch camaraderie with the South Wales miners in the deep, fiery and dangerous pits.

A model of a horse pump. A horse pump based on the Whim Gin was used underground for pumping minewater. A Whim Gin is a drum and rope used in early collieries for winding men, minerals and supplies etc. in the pit shaft. The axis of the drum was vertical and the drum was set at a height, which left room for the horse to walk a circular track around the drum, rotating it by means of a projecting beam to which the horse was attached.

Opposite, below: Baldwins Level, Pontypool, Afon Lwyd Valley, Monmouthshire in 1910. Naked lights (candles) can be seen on the left and right in the photograph. The collier on the left is cleaning up, the collier in the middle is preparing to drill a shot hole and the collier on the right is pulling the loose coal. Baldwins Level had a chequered history of openings and closures.

Committee of Monmouthshire Colliery Owners Rescue Association at the turn of the twentieth century. 'As we leave the valleys of the South Wales Coalfield it is with the realisation that we leave only physically, for there is always an aching and a longing for the valleys in our hearts.' Saint Barbara, patron Saint of miners.

Mother and Father Christmas underground giving out Black Gold – Aur Du. Coal is traditionally given for Good Luck in the New Year. The exact origin of this tradition is unknown although some believe it goes back to Medieval times when coal, bread and a candle were given to represent the fundamentals of life – food, warmth and light.

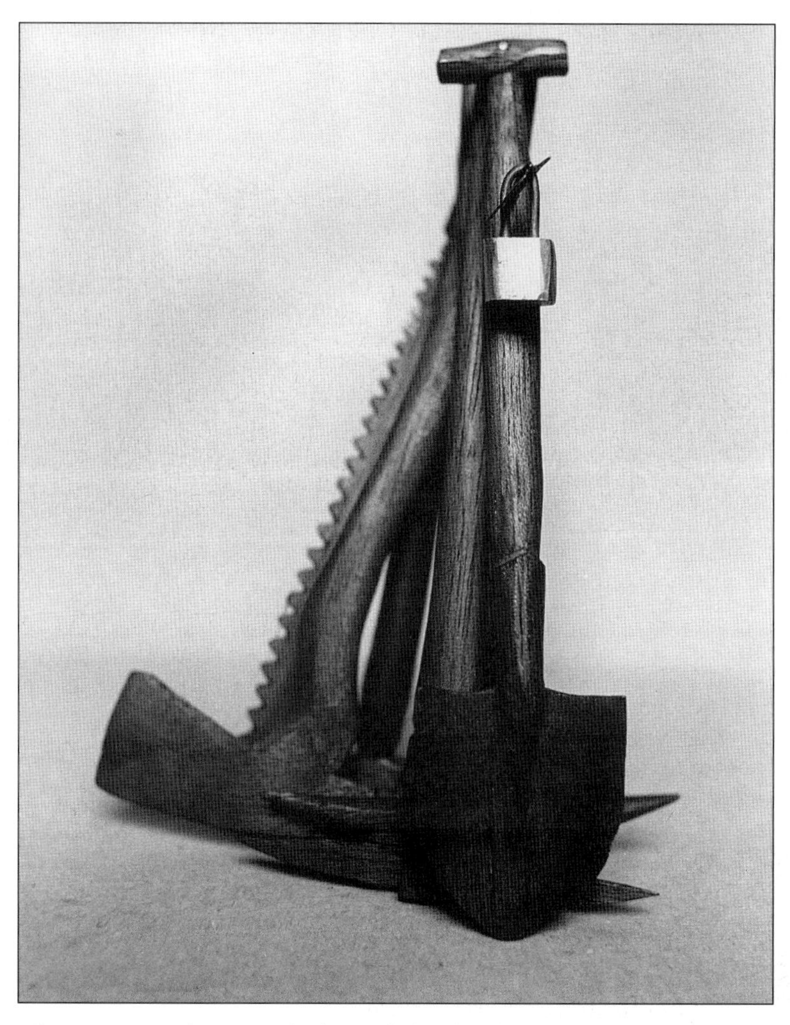

The miners' tools are on the bar – the end of the shift for the Gladiators of coalmining in the South Wales Coalfield.

A sudden change; at God's command they fell; They had no chance to bid their friends farewell, Swift came the blast, without a warning given, And bid them haste to meet their God in Heaven.

Mighty industries come and go but Mother Nature ultimately prevails and we are left with the memories of human toil and the close knit communities which are their legacy.

In Everlasting Memory to the Miners who lost their lives from Blaenavon to Newport

Date	Mine	Lives Lost
16 August 1849	Cwmynantddu, Pontypool	Five miners killed by an explosion.
19 February 1868	New Pit, Blaenavon	Roadman William Williams (25) killed by a stone which fell from the roof whilst he was clearing a fall on the main level. The cause can only be attributed to two slips meeting together in a point or angle, and thus the mass was enabled to detach itself.
24 March 1868	Old Coal Pit, Blaenavon	Collier David Jenkins (25) killed.
21 August 1872	Blaenavon	Pumpsman Thomas Tovey (57) killed, found crushed.
16 January 1886	Milfraen, Blaenavon	Sinker John Prisk (48) was injured by a stone falling and died on the 26 January. He was the chargeman of a shift of sinkers and was working in the bottom when a small stone fell on him. He had shortly before examined the pit.
21 October 1886	Dodd's Slope, Blaenavon	Collier Thomas Payne (33) killed by fall of stone. He and two others, after taking out two props, tried to bar down some stone; they got down one end but the other would not come; slips were visible in it but, nevertheless, he went underneath it to move some rubbish and it fell upon him.
9 August 1888	Garn, Blaenavon	Smithy Striker John Davis (21). In the smithy, through the bursting of a short iron tube made into a roller for the rope on an engine plane, to work over. There were cast-iron ends in this roller for the spindles or axles, one of which required repairs. It was put into the smith's fire and in five or six minutes it burst and killed this man.
15 August 1889	Forge Slope, Blaenavon	Hitcher Nat Ralph (44) killed when leaving work at the end of the day, he was riding up the engine plane on a coupling between the drams and falling off, the drams passed over him. It was contrary to orders for him to ride.
8 January 1889	Kay's Slope, Blaenavon	Haulier Henry Gibbs (26) killed by a fall of roof, stone and timber in consequence of a dram getting off the rails at a turn and knocking out a prop.

Once again a sad reminder of the true price of coal.